Summability Calculus

Ibrahim M. Alabdulmohsin

Summability Calculus

A Comprehensive Theory of
Fractional Finite Sums

 Springer

Ibrahim M. Alabdulmohsin
King Abdullah University of Science
and Technology
Dhahran, Saudi Arabia

ISBN 978-3-319-74647-0   ISBN 978-3-319-74648-7   (eBook)
https://doi.org/10.1007/978-3-319-74648-7

Library of Congress Control Number: 2018932372

Mathematics Subject Classification (2010): 11M06; 26A06; 26A24; 26A36; 33B15; 34M30; 40C05; 40G05; 41A10; 41A60; 65B10; 65B15; 65D30; 65B99; 74S20; 11B68

Printed on acid-free paper

This Springer imprint is published by Springer Nature
The registered company is Springer International Publishing AG
The registered company address is: Gewerbestrasse 11, 6330 Cham, Switzerland

To my parents,
for instilling the love of learning

Preface

This monograph is about "summability calculus," which is a comprehensive theory of *fractional* finite sums. Suppose we have a finite sum of the form $f(n) = \sum_{k=0}^{n-1} s_k\, g(k, n)$, where $g(k, n)$ is an arbitrary analytic function and s_k is some periodic sequence. Then, the claim of this monograph is that $f(n)$ is itself in an analytic form. Not only can we differentiate and integrate with respect to the bound n without having to rely on an explicit analytic formula for the finite sum, but we can also deduce asymptotic expansions, accelerate convergence, assign natural values to divergent series, and evaluate the finite sum for any $n \in \mathbb{C}$. This holds because the discrete definition of $\sum_{k=0}^{n-1} s_k\, g(k, n)$ implies a *unique* natural generalization to all $n \in \mathbb{C}$.

It will be shown that summability calculus unifies many diverse results in the literature, and that it has many important applications. For instance, it is deeply connected to the theory of summability of divergent series, which has found applications in Fourier analysis, quantum field theory, and dynamical systems, among others. Summability calculus is also connected to the finite difference methods, which are widely applied in numerical analysis and in computational science and engineering. It also contributes to, and benefits from, the methods of accelerating series convergence. Furthermore, it is intimately related to approximation methods, asymptotic analysis, and numerical integration. We will even connect summability calculus to the recent fruitful field of information theory, and use it to prove a stronger version of the celebrated Shannon-Nyquist sampling theorem.

Throughout this monograph, many celebrated results are derived, strengthened, and generalized. The list includes, but is not limited to, the Bohr-Mollerup theorem, the Stirling approximation, the Glaisher approximation, the Shannon-Nyquist sampling theorem, the Euler-Maclaurin summation formula, and the Boole summation formula. It must be noted, however, that since the author is not a professional historian himself, the various historical references presented in this monograph are only tentative, and they are mostly obtained from secondary sources.

In addition to unifying and extending disparate historical results, summability calculus can be used to derive many new aesthetic identities. For instance, by applying the "differentiation rule" of simple finite sums on the function $\sum_{k=0}^{n-1} \frac{k+1}{y+k+1}$, we prove in Eq. 2.3.19 the following identity that relates the Euler constant γ with the so-called alternating Euler constant $\log \frac{4}{\pi}$:

$$\sum_{r=1}^{\infty}\sum_{k=0}^{\infty}(-1)^k\left[\frac{1}{r+k}-\log\left(1+\frac{1}{r+k}\right)\right]=\frac{\log\frac{4}{\pi}+\gamma}{2}$$

As a second example, we can use summability calculus to derive closed-form expressions related to the sums of approximation errors of the harmonic numbers (see Theorem A.1 in Appendix A):

$$\sum_{k=1}^{\infty}\left[\log(k)+\gamma-H_k+\frac{1}{2k}\right]=\frac{\log(2\pi)-1-\gamma}{2}$$

$$\sum_{k=1}^{\infty}\left[\log\sqrt{k(k+1)}+\gamma-H_k\right]=\frac{\log(2\pi)-1}{2}-\gamma$$

$$\sum_{k=1}^{\infty}(-1)^k\left[\log k+\gamma-H_k\right]=\frac{\log\pi-\gamma}{2}$$

These identities are proved using tools that are available in Chaps. 5 and 6, which, in turn, rely on foundational results in Chaps. 2 and 4. Remarkably, the error terms of approximating the harmonic numbers using the logarithmic function and the Euler constant contain the Euler constant themselves!

Moreover, we show how summability calculus can be used to derive identities related to the Gregory coefficients and the Cauchy coefficients in Chap. 7. Writing G_k for the Gregory coefficients, we show in Sect. 7.3.3 that:

$$|G_1|+|G_2|-|G_4|-|G_5|+|G_7|+|G_8|-|G_{10}|-|G_{11}|+\cdots=\frac{\sqrt{3}}{\pi}$$

$$|G_1|-|G_3|-|G_4|+|G_6|+|G_7|-|G_9|-|G_{10}|+|G_{12}|+\cdots=1-\frac{\sqrt{3}}{\pi}$$

Similar identities for the Cauchy coefficients are proved as well.

Although we opted to illustrate some of the new identities here in this preface, the real contribution of this monograph is to provide a comprehensive theory for fractional finite sums. This includes methods for differentiating, integrating, and computing asymptotic expansions to fractional finite sums. It also includes methods for evaluating fractional finite sums that go beyond polynomial approximation. We integrate all of these results with the theory of summability of divergent series as well. In fact, we also present a new summability method in this monograph in Chap. 4 and prove some of its convenient properties.

Dhahran, Saudi Arabia

Ibrahim M. Alabdulmohsin

November 2017

Contents

Chapter 1
Introduction

One should always generalize

Carl Jacobi (1804–1851)

Abstract Generalization has been an often-pursued goal in mathematics. We can define it as the process of introducing new systems and operations in order to extend the domains of existing systems and operations consistently, while still preserving as many prior results as possible. In this chapter, we look into the question of how to generalize discrete finite sums and products from the set of integers into the entire complex plane, which give rise to the notion of fractional finite sums. We discuss the history of this problem in this chapter, describe the contributions of this book, and sketch an outline to the remainder of the chapters.

1.1 Preliminaries

1.1.1 Generalization

Generalization has been an often-pursued goal in mathematics. We can define it as the process of introducing *new* systems and operations in order to extend the domains of *existing* systems and operations consistently, while still *preserving* as many prior results as possible. Generalization is constantly pursued in many areas of mathematics, including fundamental concepts such as numbers and geometry, systems of operations such as the arithmetic, and the domains of functions using analytic continuation.

One historical example of a mathematical generalization that is of particular interest in this monograph is extending the domains of special discrete functions, such as finite sums and products, to the complex plane $\mathbb{C}$. Consider, for instance, the following sequences:

The Factorials:	$1, 1, 2, 6, 24, 120, \ldots$
The Triangular Numbers:	$1, 3, 6, 10, 15, 21, 28, \ldots$

© Springer International Publishing AG 2018
I. M. Alabdulmohsin, *Summability Calculus*,
https://doi.org/10.1007/978-3-319-74648-7_1

The Harmonic Numbers: $1, \dfrac{3}{2}, \dfrac{11}{6}, \dfrac{25}{12}, \dfrac{137}{60}, \dfrac{49}{20}, \ldots$

The Zeta Sequence: $1, \dfrac{5}{4}, \dfrac{251}{216}, \dfrac{726}{673}, \dfrac{820}{791}, \dfrac{822}{802}, \ldots$

The Alternating Integers: $1, -1, 2, -2, 3, -3, \ldots$

For each of the sequences above, one can find a smooth analytic function $f(n) : \mathbb{C} \to \mathbb{C}$ that perfectly fits the whole sequence. In fact, it is even entirely possible to find *infinitely many* functions. However, such a "blind" fitting is not very meaningful in extending the definition of discrete sequences to non-integer arguments. What one truly desires is not an interpolation per se, but, rather, a *generalization*. In informal terms, generalization not only aims at fitting a sequence, but it also aims at *preserving* some additional fundamental properties as well. Consequently, generalization has an intrinsic significance that provides a deeper insight, and leads naturally to an evolution of mathematical thought.

Looking into the first three sequences above, we note that the factorial function satisfies the recurrence identity:

$$f(n) = n \cdot f(n-1), \tag{1.1.1}$$

the triangular numbers satisfy the identity:

$$f(n) = n + f(n-1), \tag{1.1.2}$$

and the harmonic numbers satisfy the identity:

$$f(n) = \frac{1}{n} + f(n-1). \tag{1.1.3}$$

These identities state that the log-factorial function $\log n!$ as well as the triangular and harmonic numbers are *finite sums* of the form $\sum_{k=0}^{n-1} g(k)$ for some analytic function $g(k)$. Keeping these properties in mind, one can now look for a "natural" generalized definitions $f_G(n) : \mathbb{C} \to \mathbb{C}$ to such finite sums that preserve the aforementioned recurrence identities.

More precisely, suppose we have[1]:

$$f(n) = \sum_{k=0}^{n-1} g(k).$$

---

[1]We will use this convention throughout the monograph because it leads to more elegant mathematics than, for instance, $f(n) = \sum_{k=a}^{n} g(k)$. Of course, this does not limit the generality of our results.

Then, the following question arises: Is there a unique natural method of extending the domain of $f(n)$ to the complex plane $\mathbb{C}^2$? If such a generalized definition exists, and indeed it always does, can we look for a similar generalization to finite sums of the, somewhat more complex, form $\sum_{k=0}^{n-1}(1+k)^{-n}$, which produces the earlier "zeta sequence", and for oscillating sums of the form $\sum_{k=0}^{n-1}(-1)^{1+k}(1+k)$, which produces the "alternating integers" sequence?

More generally, suppose we have a finite sum of the form:

$$f(n) = \sum_{k=0}^{n-1} s_k\, g(k, n), \qquad (1.1.4)$$

where $g(k, n)$ can depend on *both* the iterated variable k and the bound n, while s_k is some periodic sequence. Then, the question remains: Is there a unique natural method of extending the domain of $f(n)$ in Eq. 1.1.4 to the complex plane $\mathbb{C}$? Surprisingly, we will show in this monograph that the answer to the latter question is in the affirmative.

1.1.2 An Example: The Power Sum Function

To keep our discussion concrete, let us focus our attention, first, on the discrete *power sum* function $f(\cdot\,; m) : \mathbb{N} \to \mathbb{N}$, of which triangular numbers are a particular case:

$$f(n; m) = \sum_{k=0}^{n-1}(1+k)^m. \qquad (1.1.5)$$

In Eq. 1.1.5, it is trivial to realize that an infinite number of smooth analytic functions can correctly interpolate $f(\cdot\,; m)$. In fact, let $f_G : \mathbb{C} \to \mathbb{C}$ be a valid function, i.e. $f_G(n) = f(n; m)$ for all $n \in \mathbb{N}$. Then, adding $f_G(n)$ to some periodic function $h : \mathbb{C} \to \mathbb{C}$ that satisfies $h(n) = 0$ for all $n \in \mathbb{N}$ will also provide us with an alternative definition of $f(n; m)$ for all $n \in \mathbb{C}$. For example, all of the following functions $f_G(n; k)$ for different choices of $k \in \mathbb{Z}$ correctly fit the *triangular numbers* mentioned earlier[3]:

$$f_G(n; k) = \frac{n\,(n+1)}{2} + \sin(k\,\pi\,n) \qquad \text{for any } k \in \mathbb{Z}.$$

---

[2]We ignore the issue of singularities here that may arise when extending the domain to the complex plane $\mathbb{C}$. Because the domain of $f(n)$ is extended in a unique natural manner, those singularities are unavoidable. In fact, they can be viewed as quite natural. For example, consider the factorial function $f(n) = \prod_{k=0}^{n-1}(k+1)$. We have by definition $f(1) = 1$ and by the recurrence identity $f(1) = 1 \cdot f(0)$. Thus, $f(0) = 1$. However, by the same recurrence identity $f(0) = 1 = 0 \cdot f(-1)$, which implies that the factorial function must have a singularity at $n = -1$.

[3]To recall, the triangular numbers are obtained from Eq. 1.1.5 by setting $m = 2$.

However, not all $f_G(n; k)$ satisfy the triangular numbers' recurrence identity $f(n) = n + f(n-1)$, and among those that do, only the function $f_G(n; 0)$ is a polynomial. Such a polynomial function $f_G(n; 0)$ is a particular case of a more general solution, called the *Bernoulli-Faulhaber formula*.

The Bernoulli-Faulhaber formula yields a polynomial that preserves the recurrence identity of $f(n; m)$ that is given in Eq. 1.1.6 for all $n \in \mathbb{C}$, which makes it a suitable candidate for a generalized definition of power sums. In fact, it is indeed the *unique* family of *polynomials* that enjoys such an advantage. Hence, it is the unique most natural generalized definition of power sums if one considers polynomials to be the "simplest" of all possible functions. The Bernoulli-Faulhaber formula is given in Eq. 1.1.7, where B_r are Bernoulli numbers and $B_1 = -\frac{1}{2}$.[4]

$$f(n; m) = n^m + f_m(n-1) \tag{1.1.6}$$

$$f(n; m) = \frac{1}{m+1} \sum_{j=0}^{m} (-1)^j \binom{m+1}{j} B_j\, n^{m+1-j}. \tag{1.1.7}$$

If a discrete finite sum can be interpolated by a polynomial that also happens to satisfy the required recurrence identity, then that polynomial is, perhaps, the unique most natural method of extending the domain of the discrete finite sum. However, not all finite sums are interpolable with polynomials. One famous example is the log-factorial function. Because the factorials are encountered in many areas of mathematics including combinatorics and analysis, extending the domain of the log-factorials function $\sum_{k=0}^{n-1} \log(1 + k)$ is an important feat, since this is equivalent to the task of extending the domain of the factorial function itself. Despite its popularity, this problem had withstood many unsuccessful attempts by eminent mathematicians, such as Bernoulli and Stirling, until Euler came up with his famous answer in the early 1730s in a series of letters to Goldbach that introduced his infinite product formula and the gamma function [Eul38b, Dav59, Weib]. Indeed, arriving at the gamma function from the discrete definition of the factorials was not a simple task, needless to mention proving that it was the *unique* natural generalization of the factorials, as the Bohr-Mollerup theorem arguably stated [Kra99].

How do we proceed systematically from the discrete definition of the factorial function to the gamma function? And how do we know that the gamma function is indeed the *unique* most natural method of generalizing the definition of the factorials? Clearly, a comprehensive theory is needed for answering these questions. We hope that by providing a comprehensive formal theory for answering these

---

[4]The reader is kindly referred to the Glossary section for a definition of Bernoulli numbers.

questions, one will be able to apply it readily to the potentially infinite list of special discrete functions, such as the factorial-like *hyperfactorial* and *superfactorial* functions defined in Eqs. 1.1.8 and 1.1.9 respectively.[5]

$$\textbf{Hyperfactorial:} \quad H(n) = \prod_{k=1}^{n} k^k \qquad (1.1.8)$$

$$\textbf{Superfactorial:} \quad S(n) = \prod_{k=1}^{n} k! \qquad (1.1.9)$$

1.1.3   Goals

Aside from extending the domains of discrete finite sums, performing infinitesimal calculus is a second goal that we would like to achieve. This includes being able to differentiable and integrate a finite sum $\sum_{k=0}^{n-1} g(k)$ with respect to its argument n. This second objective of performing infinitesimal calculus is intimately connected to our previous objective of extending the domain of the function. However, the two objectives are not equivalent.

Consider the following function:

$$H_n = \sum_{k=0}^{n-1} \frac{1}{1+k}. \qquad (1.1.10)$$

The function H_n is commonly referred to as the *harmonic number*, whose first few values were given earlier in our third number sequence example. The first goal we mentioned previously was to deduce a natural definition of H_n for all complex values of n so that a finite sum such as $\sum_{k=1}^{1/2} \frac{1}{k}$ becomes meaningful. As will be revealed in this monograph, providing such a definition can be achieved using a series representation of H_n, which follows *immediately* from the *discrete* definition of H_n. This series representation yields $H_{1/2} = 2(1 - \log 2)$.

Computing a function's derivative, on the other hand, can certainly utilize the series representation of H_n mentioned above, but it does not need to. In fact, we will show that we can compute all of the derivatives $\frac{d^r}{dn^r} H_n$ at $n \in \mathbb{N}$ for all $r \geq 0$ *without* knowing how to define H_n for $n \in \mathbb{C}$ in the first place! The key insight towards achieving such a goal is to impose some "natural" constraints on any generalized definition of H_n. In our example, we note that $(H_n - H_{n-1}) \to 0$ as $n \to \infty$ so we impose the constraint that $\frac{d}{dn} H_n \to 0$ as $n \to \infty$. This immediately "proves" that the derivative $\frac{d}{dn} H_n$ at $n = 0$ is $\zeta(2)$, where $\zeta(s)$ is the Riemann zeta function.

---

[5]The reader may refer to [Weid] and [Weia] for a brief introduction to such family of functions.

How do we formalize and carry out such a process? The details will be given later in Chap. 2.

Our third fundamental goal is to be able to determine the asymptotic behavior of finite sums. Without a doubt, asymptotic analysis has profound applications. For example, Euler showed that the harmonic numbers were asymptotically related to the logarithmic function, which led him to introduce the famous constant γ, defined by Eq. 1.1.11 [Eul38a, San07]. Stirling, on the other hand, presented the famous asymptotic formula for the factorial function given in Eq. 1.1.12, which is used ubiquitously, such as in the study of algorithms and data structures [Rob55, LRSC01]. Additionally, Glaisher in 1878 presented an asymptotic formula for the hyperfactorial function given in Eq. 1.1.13, which led him to introduce a new constant, denoted A in Eq. 1.1.13, that is intimately related to Euler's constant and the Riemann zeta function [Gla93, Weic, Weid]. The constant A is sometimes referred to as the Kinkelin constant [CS97].

$$\lim_{n \to \infty} \left\{ \sum_{k=0}^{n-1} \frac{1}{1+k} - \log n \right\} = \gamma \tag{1.1.11}$$

$$n! \sim \sqrt{2\pi n} \left(\frac{n}{e} \right)^n \tag{1.1.12}$$

$$H(n) \sim A \, n^{\frac{n^2+n}{2} + \frac{1}{12}} e^{-\frac{n^2}{4}}, \quad A \approx 1.2824 \tag{1.1.13}$$

While these results have been deduced at different periods of time using different approaches, a fundamental question that arises is whether there exists a comprehensive, unifying theory that bridges all of these different results together and makes their proofs nearly elementary. More crucially, we desire that such a theory yields a simple approach for determining the asymptotic behavior of the more general family of finite sums of the form $\sum_{k=0}^{n-1} s_k g(k, n)$, where, again, s_k is an arbitrary periodic sequence.

So, to summarize, we have three fundamental goals that we would like to achieve:

1. To extend the domain of discrete finite sums to the complex plane $\mathbb{C}$.
2. To be able to perform infinitesimal calculus on discrete finite sums, such as differentiation, integration, and Taylor series expansion.
3. To determine the asymptotic behavior of discrete finite sums.

Not surprisingly, all of the three goals are closely connected to each other. Quite surprisingly, they are occasionally connected via seemingly unrelated topics, such as the calculus of finite differences, the theory of summability of divergent series, and to methods of accelerating series convergence. In general, these three goals can be achieved using what we will refer to from now on as "summability calculus".

1.2   What is Summability Calculus?

Summability calculus can be defined in a nutshell as follows. Given a discrete finite sum of the form $\sum_{k=0}^{n-1} s_k g(k, n)$, where $g(k, n)$ is an analytic function and s_k is a periodic sequence, then the finite sum is in an analytic form. Not only can we perform differentiation and integration with respect to n without having to rely on an explicit analytic formula for the finite sum, but we can also immediately evaluate the finite sum for fractional values of n, deduce asymptotic expressions, accelerate convergence of the infinite sum, assign natural values to divergent series, and come up with a potentially infinite list of interesting identities as a result; all without having to *explicitly* extend the domain of $f(n)$ to the complex plane $\mathbb{C}$. To reiterate, this follows because the expression $\sum_{k=0}^{n-1} s_k g(k, n)$ embodies within its discrete definition an immediate natural definition of $f(n)$ for all $n \in \mathbb{C}$.

1.2.1   Duality

Summability calculus vs. conventional infinitesimal calculus can be viewed in light of an interesting duality. In infinitesimal calculus, the infinitesimal behavior at a *local* interval is used to determine the *global* behavior at the entire analytic disc. For instance, the Taylor series expansion of a function $f(x)$ around a point $x = x_0$ is often used to compute the value of $f(x)$ for some points x outside x_0, sometimes even in the entire function's domain. This is illustrated in Fig. 1.1. Thus, the behavior of the analytic function at an infinitesimally small interval is sufficient to deduce the

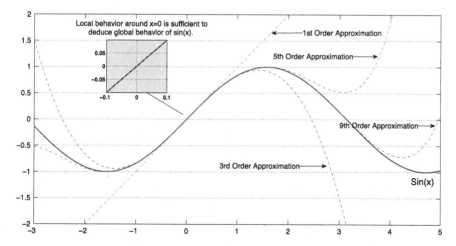

Fig. 1.1 In infinitesimal calculus, the local behavior of a function is used to reconstruct its global behavior. In this figure, for example, the Taylor series approximations around the origin $x = 0$ is used to reconstruct the function $\sin x$

global behavior of the function. Such an incredible property of analytic functions, often referred to as *rigidity*, is perhaps the cornerstone of infinitesimal calculus that has led its many wonders.

In summability calculus, on the other hand, we follow the contrary approach by employing our incomplete knowledge about the global behavior of a function to reconstruct accurately its local behavior at any desired interval. Consider the example of a geometric finite sum $f(n) = \sum_{k=0}^{n-1} x^k$ for some $|x| < 1$. Such a finite sum has a well-known closed-form formula. However, suppose that we do not know what that closed-from formula is but we know that the limit $\lim_{n \to \infty} f(n)$ exists. Using such a limited knowledge, we would expect $f'(n) \to 0$ as $n \to \infty$. It will be shown later in Chap. 2 that the derivative of a finite sum is always given by the form:

$$\frac{d}{dn} \sum_{k=0}^{n-1} g(k) = c + \sum_{k=0}^{n-1} g'(k).$$

Here, c is a *definite*, non-arbitrary constant. To determine the value of c for the geometric sum $f(n) = \sum_{k=0}^{n-1} x^k$, we use our knowledge that $f'(n) \to 0$ as $n \to \infty$. This gives us:

$$c = -\log x \sum_{k=0}^{\infty} x^k = -\frac{\log x}{1-x}.$$

We will also show in Chap. 2 that c is always equal to $f'(0)$. Putting both facts together, we deduce that: $f'(0) = -\frac{\log x}{1-x}$. Needless to mention, this result indeed holds as one can easily verify using the closed-form expression of the geometric sum. For now, it is worthwhile to note that we have used our knowledge about the asymptotic behavior of $f(n)$ to compute the derivative $f'(0)$ at the origin.

1.2.2 Sampling and Interpolation

The duality above brings to mind the well-known Shannon-Nyquist sampling theorem, which is one of the fundamental results in information theory. Here, if a function is *bandlimited*, meaning that its non-zero frequency components are restricted to a bounded region in the frequency domain, then discrete samples taken at a sufficiently high rate can be used to reconstruct the function completely *without any loss of information*. In this case, our global knowledge about the discrete samples can be used to determine the behavior of the function at all intervals, which is similar to what summability calculus fundamentally entails. Does summability calculus have anything to do with the sampling theorem? Surprisingly, the answer is yes. In fact, we will use some of the results of summability calculus to prove a slightly stronger version of the Shannon-Nyquist sampling theorem.

Sampling necessitates interpolation. One crucial link between interpolation and summability calculus is polynomial fitting. For example, given discrete samples of the power sum function mentioned earlier in Eq. 1.1.5, then such samples can be perfectly interpolated with polynomials. Assuming that polynomials are the simplest of all possible functions, these interpolating polynomials are, by this measure, the *unique most natural generalization* to power sums. In summability calculus, polynomial fitting is a cornerstone. It is what guarantees summability calculus to operate on the unique most natural generalization to finite sums; that is, even if the finite sums themselves cannot be fitted with polynomials! Such a claim might appear, at first sight, to be paradoxical; if the harmonic numbers, the log-factorial function, the zeta sequence, and the alternating integers sequence presented earlier could not be fitted by polynomials, why would their fractional values be intimately related to polynomial fitting? The answer to this question will become clear later in Chaps. 2 and 6.

1.3   What Are Fractional Finite Sums?

We have stated that summability calculus allows us to perform infinitesimal calculus, such as differentiation and integration, and to evaluate fractional finite sums without having to explicitly extend the domains of finite sums to the complex plane $\mathbb{C}$. However, if a finite sum is discrete in nature, what does its fractional values mean in the first place and why are they important? To answer these questions, we need to digress to a more fundamental question: what does a finite sum of the form $f(n) = \sum_{k=0}^{n-1} g(k)$ fundamentally represent?

Traditionally, the symbol $\sum$ was meant to be used as a shorthand notation for an iterated addition, and was naturally restricted to discrete values. However, it turns out that if we *define* a finite sum more broadly by *two additional properties*, we can deduce a natural definition that extends the domain of $f(n)$ to the complex plane $\mathbb{C}$. The two properties are:

1. $\forall x, g(\cdot) : \displaystyle\sum_{k=x}^{x} g(k) = g(x)$

2. $\forall a, b, x, g(\cdot) : \displaystyle\sum_{k=a}^{b} g(k) + \sum_{k=b+1}^{x} g(k) = \sum_{k=a}^{x} g(k).$

These properties are two of six axioms proposed by Müller and Schleicher in 2011 for defining fractional finite sums [MS11]. Their work is restricted to what is referred to in this monograph as *simple* finite sums of *finite polynomial order*. In this monograph, however, we will not pursue an axiomatic treatment. Instead, we will show that a definition of fractional finite sums arises naturally" out of those two properties alone. In fact, we will show that fractional finite sums can be defined in a much broader setting than the one considered in [MS11], such as *composite* and *oscillatory* finite sums.

First, we note that the two properties above suffice to recover the original discrete definition of finite sums when $n \in \mathbb{N}$ as follows:

$$\sum_{k=0}^{n-1} g(k) = \sum_{k=0}^{n-2} g(k) + \sum_{k=n-1}^{n-1} g(k) \qquad \text{(by property 2)}$$

$$= \sum_{k=0}^{n-3} g(k) + \sum_{k=n-2}^{n-2} g(k) + \sum_{k=n-1}^{n-1} g(k) \qquad \text{(by property 2)}$$

$$= \sum_{k=0}^{0} g(k) + \sum_{k=1}^{1} g(k) + \cdots + \sum_{k=n-1}^{n-1} g(k) \qquad \text{(by property 2)}$$

$$= g(0) + g(1) + \cdots + g(n-1). \qquad \text{(by property 1)}$$

In addition, we recover the recurrence identity given in Eq. 1.3.1. This recurrence identity is extremely important in subsequent analysis.

$$f(n) = \sum_{k=0}^{n-1} g(k) = \sum_{k=n-1}^{n-1} g(k) + \sum_{k=0}^{n-2} g(k) = g(n-1) + f(n-1). \qquad (1.3.1)$$

Moreover, we immediately observe that the two defining properties of finite sums imply a *unique* natural generalization to all $n \in \mathbb{C}$ *if the infinite sums exist*. By property 2, we have:

$$\sum_{k=0}^{n-1} g(k) = \sum_{k=0}^{\infty} g(k) - \sum_{k=n}^{\infty} g(k). \qquad (1.3.2)$$

Therefore, if the two infinite sums are well-defined for $n \in \mathbb{C}$, the finite sum $\sum_{k=0}^{n-1} g(k)$ is *uniquely* defined as well.

Consequently, it seems obvious that a unique most natural generalization can be defined if the infinite sums in Eq. 1.3.2 both exist. At least, this is what we would expect if the term "natural generalization" has any meaning. What is not obvious, however, is that a *natural* generalization of finite sums is *uniquely* defined for all n and all analytic functions g. In fact, even a finite sum of the form $\sum_{k=0}^{n-1} s_k g(k, n)$ has a unique natural generalization that extends its domain to the complex plane $\mathbb{C}$. However, the argument in the latter case is more intricate as will be revealed later.

In addition, we will also show that the study of divergent series and *analytic summability theory* is intimately tied to the definition of fractional sums. In a few words, summability theory is concerned with methods for assigning "natural" values to divergent series. For example, the *Grandi series* $1 - 1 + 1 - 1 + 1 - \ldots$ oscillates indefinitely between 0 and 1, and, hence, it does not have a well-defined value in the strict sense of the word. Despite this, one can argue that the Grandi series is *equivalent* to $1/2$ using countless arguments. One argument is Bernoulli's frequentist interpretation, which suggests that since the partial sum equals to zero half of the time and equals to one half of the time, its *expected* value is $1/2$. A second argument is the Abel summability method, which appeals to continuity as a method for defining divergent series and gives the same value of $1/2$. A third argument is using the Taylor series expansion of $(1 + x)^{-1}$, which again gives the same value of $1/2$, and so on.

Summability of divergent series is intimately related to fractional sums. To see this, let us return to our earlier example of the alternating integer sequence given by:

$$f(n) = \sum_{k=0}^{n-1} (-1)^{k+1} k$$

In order to compute $f(\frac{3}{2})$, we will use Eq. 1.3.2 and write it as:

$$\sum_{k=0}^{\frac{1}{2}} (-1)^{k+1} k = \sum_{k=0}^{\infty} (-1)^{k+1} k - \sum_{k=\frac{3}{2}}^{\infty} (-1)^{k+1} k$$

Using summability methods such as the Cesàro means, we have $\sum_{k=0}^{\infty} (-1)^{k+1} k = \frac{1}{4}$ and $\sum_{k=\frac{3}{2}}^{\infty} (-1)^{k+1} k = i\frac{1}{2}$, where i is the complex number. Using both values, we deduce that $\sum_{k=0}^{1/2} = \frac{1}{4} - i\frac{1}{2}$. Now, it is straightforward to observe that $f(n)$ has, in fact, the following closed-form expression:

$$\sum_{k=0}^{n} (-1)^{k+1} g(k) = \frac{1}{4} - (-1)^n \frac{2n+1}{4}.$$

Plugging $n = 1/2$ into the above expression yields the same value $\sum_{k=0}^{1/2} = \frac{1}{4} - i\frac{1}{2}$.

How to formalize such a process and prove its correctness is the subject of Chaps. 4 and 5. For now, we will only say that a generalized definition of series exists, denoted $\mathfrak{T}$, which is consistent with the traditional definition of summation. Using $\mathfrak{T}$, we will show how the study of oscillating sums is greatly simplified. For example, if a divergent series $\sum_{0}^{\infty} g(k)$ is *summable* by $\mathfrak{T}$, then, as far as summability calculus is concerned, the series behaves *exactly as if it were convergent*! Here, the term "summable" divergent series, loosely speaking, means that a natural value can be assigned to the divergent series using the generalized definition $\mathfrak{T}$. For example, one special case of $\mathfrak{T}$ is the Abel summability method, which assigns to a divergent series $\sum_{0}^{\infty} g(k)$ the value $\lim_{z \to 1^-} \sum_{k=0}^{\infty} g(k) z^k$ if the limit exists. Intuitively speaking, this method appeals to continuity as a rational basis for defining divergent series, which is similar to the arguments that $\sin z/z = 1$ and $z \log z = 0$ at $z = 0$. The Abel summability method was used prior to Abel. In fact, Euler used it quite extensively and called it the "generating function method" [Var07]. In addition, Poisson also used it frequently in summing Fourier series [Har49].

Using the generalized definition of series given by $\mathfrak{T}$, we will show that summability theory generalizes the earlier statement that a finite sum is naturally defined by Eq. 1.3.2, which otherwise would only be valid if the infinite sums converge. Using $\mathfrak{T}$, on the other hand, we will show that the latter equation holds, in general, if the infinite sums exist in $\mathfrak{T}$, i.e. even if they are not necessarily convergent in the classical sense. However, not all divergent series are defined in $\mathfrak{T}$ so a more general statement will also be presented.

With this in mind, we are now ready to see what a fractional finite sum means. In brief terms, since we are claiming that any finite sum implies a *unique* natural generalization that extends its domain from a discrete set of numbers to the complex plane $\mathbb{C}$, the value of a fractional sum is equivalent to the value of its unique natural generalization. The key advantage of summability calculus is that we can compute fractional finite sums and perform infinitesimal calculus *without* having to determine an explicit analytic expression in the classical sense of the word. So, notations such as $\frac{d}{dt} \sum'$ and even $\int \sum' dt$ will become, not only meaningful, but quite common as well.

In summary, this monograph presents the foundations of summability calculus, which bridges the gap between various well-established results in number theory, summability theory, infinitesimal calculus, and the calculus of finite differences. In fact, and as will be shown throughout this monograph, it contributes to more branches of mathematics such as approximation theory, asymptotic analysis, information theory, and the study of accelerating series convergence. However, before we begin discussing summability calculus, we first have to outline some terminologies that will be used quite frequently in the sequel.

1.4 Terminology and Notation

$$f(n) = \sum_{k=0}^{n-1} g(k), \qquad \text{(SIMPLE)}$$

$$f(n) = \sum_{k=0}^{n-1} g(k, n), \qquad \text{(COMPOSITE)}$$

$$f(n) = \sum_{k=0}^{n-1} s_k\, g(k), \qquad \text{(OSCILLATORY SIMPLE)}$$

$$f(n) = \sum_{k=0}^{n-1} s_k\, g(k, n), \qquad \text{(OSCILLATORY COMPOSITE)}$$

Simple finite sums are defined in this monograph to be any finite sum given by the general form shown above, where k is an iteration variable. In a composite sum, on the other hand, the iterated function g depends on *both* the iteration variable k and the bound n as well. Clearly, simple finite sums are, themselves, composite finite sums. In addition, if a periodic sequence s_k is present, we will refer to simple or composite finite sums as *oscillatory* simple or competitive finite sums, respectively.

Clearly, simple finite sums are a particular case of oscillatory simple finite sums, where $s_k = 1$ for all $k \geq 0$.

To facilitate readability, we will strive to simplify mathematical notation as much as possible. For instance, we will typically use $f(n)$ to denote a discrete function and let $f_G(n) : \mathbb{C} \to \mathbb{C}$ denotes its unique natural generalization in almost every example. Here, the notation $\mathbb{C} \to \mathbb{C}$ is meant to imply that both the domain and the range are simply connected regions of the complex plane. In other words, we will not explicitly state the domain of every single function since it can often be inferred without notable efforts from the function's definition. Similarly, a statement such as "f converges to g" will be used to imply that f approaches g whenever g is defined and that both share the same domain. Thus, we avoid statements such as "f converges to g in $\mathbb{C} - \{s\}$ and has a simple pole at s", when both f and g have the same poles. Whereas summability calculus extends the domain of a finite sum $\sum_{k=0}^{n-1} s_k \, g(k, n)$ from a subset of integers to the complex plane $n \in \mathbb{C}$, we will usually focus on examples for which n is real-valued.

In addition, the following important definition will be used quite extensively.[6]

Definition 1.1 (Polynomial Order) A function $g(x)$ is said to have a *finite polynomial order* if a non-negative integer m exists such that $g^{(m+1)}(x) \to 0$ as $x \to \infty$. The minimum non-negative integer m that satisfies such a condition for a function g will be called its *polynomial order*.

In other words, if a function $g(x)$ has an finite polynomial order m, then *only the function up to its mth derivative matter asymptotically*. For example, any polynomial with degree n has a polynomial order of n. A second example is the function $g(x) = \log x$, which has a polynomial order of zero because its first derivative goes to zero as $x \to \infty$. Non-polynomially bounded functions such as e^x and the factorial function $x!$ have infinite polynomial orders.

Finally, in addition to the hyperfactorial and the superfactorial functions described earlier, the following functions will be used frequently throughout this monograph:

Generalized Harmonic Numbers: $\quad H_{m,n} = \sum_{k=0}^{n-1} k^{-m}$

Gamma Function: $\quad \Gamma(n) = \int_0^\infty e^{-t} t^{n-1} \, \mathrm{d}t$

---

[6]The significance of this concept was recognized by Müller and Schleicher [MS10], who called functions that have a finite polynomial order "approximate polynomials". Independently and almost simultaneously, the same concept was used by the author in an earlier draft of this monograph [Ala12], who called it a "finite differentiation order". The work of Müller and Schleicher [MS10] is restricted to simple finite sums of a finite polynomial order, whereas the author used this concept to derive more general results that hold for oscillatory composite finite sums of arbitrary, possibly infinite, polynomial orders.

Gauss PI Function:	$\Pi(n) = \Gamma(n+1) = n!$
Log-Factorial Function:	$\varpi(n) = \log \Pi(n) = \log n!$
Digamma Function:	$\psi(n) = \frac{d}{dn} \log \Gamma(n)$
Riemann Zeta Function:	$\zeta(s) = \sum_{k=0}^{\infty} (1+k)^{-s}$, for $s > 1$.

Some of these functions, as well as other frequently used terms, are listed in the Glossary section at the end of this monograph.

Here, it is worth mentioning that the superfactorial function $S(n)$ is also often defined using the *double gamma function* or *Barnes G-function* [CS00, Weia]. All of these definitions are essentially equivalent. In this monograph, we will exclusively use the superfactorial function as defined earlier in Eq. 1.1.9. Moreover, we will almost always use the log-factorial function $\log \Gamma(n+1)$, as opposed to the log-gamma function. Legendre's normalization $n! = \Gamma(n+1)$ is unwieldy, and will be avoided to simplify mathematics. It is perhaps worthwhile to note that avoiding such a normalization of the gamma function is not a new practice. In fact, Legendre's normalization has been avoided and even harshly criticized by some twentieth century mathematicians such as Lanczos, who described it as "void of any rationality" [Lan64, Spo94].

1.5 Historical Remarks

Finite sums have a long history in mathematics. In 499 AD, the great Indian mathematician Aryabhata investigated the sum of arithmetic progressions and presented its closed-form formula in his famous book *Aryabhatiya* when he was 23 years old. In this book, he also stated formulas for the sums of powers of integers up to the summation of cubes. Unfortunately, mathematical notations were immature during that era, and the great mathematician had to state his mathematical equations using plain words: *"Diminish the given number of terms by one then divide by two then ..."* [Vol77]. In addition, the Greeks were similarly interested in finite sums as Euclid demonstrated in his Elements Book IX Proposition 35, in which he presents the well-known formula for the sum of a finite geometric progression. Between then and now, countless other mathematicians were fascinated with the subject of finite sums.

Given the great interest mathematicians have placed in summations and knowing that this monograph is all about finite sums, it should not be surprising to see that a large number of the results presented here have already been discovered at different points in time during the past 1500 years. In fact, some of what is rightfully considered as building blocks of summability calculus were deduced 300 years ago, while others were published as recently as 2010. Nonetheless, a large portion of this monograph appears to be new.

The earliest known work that is directly related to summability calculus is the Gregory quadrature formula, whose original intention was in numerical integration [Jor65, SP11]. The Gregory quadrature formula approximates definite integrals using finite sums and finite differences. We will show that Gregory's formula, when interpreted formally, indeed corresponds to the unique most natural generalization of simple finite sums. Later around the year 1735, Euler and Maclaurin independently came up with the celebrated Euler-Maclaurin summation formula that also extends the domain of simple finite sums to non-integer arguments [Apo99, San03]. The Euler-Maclaurin formula has been widely applied, and was described as one of "the most remarkable formulas of mathematics" [Lam01]. Here, it is worth mentioning that neither Euler nor Maclaurin published a formula for the remainder term, which was first developed by Poisson in 1823 [Lam01]. The difference between the Euler-Maclaurin summation formula and Gregory's is that finite differences are used in the latter formula as opposed to infinitesimal derivatives. Nevertheless, they are both formally equivalent, and one can be derived from the other [SP11]. Later, George Boole introduced his summation formula, which is the analog of the Euler-Maclaurin summation formula for alternating sums [BCM09]. The Boole summation formula agrees, at least formally, with the Euler-Maclaurin summation formula, and there is evidence that suggests it was known to Euler as well [BBD89]. In 1975, Berndt and Schoenfeld provided "periodic" analogs to the Euler-Maclaurin summation formula, which encompass the Boole summation formula as a special case [BS75].

In this monograph, we will derive these formulas, generalize them to oscillatory composite finite sums, prove their uniqueness for being the most natural generalization to finite sums, and present their counterparts using finite differences as opposed to infinitesimal derivatives. We will show that the Euler-Maclaurin summation formula arises out of polynomial fitting, and show that the Boole summation formula, as well as the Berndt-Schoenfeld periodic analogs, are intimately connected to the subject of divergent series and summability theory. In particular, we establish how the asymptotic formulas with periodic signs can also be generalized to "non-periodic signs", using the theory of summability of divergent series. Most importantly, we will show that it is more convenient to use all of these asymptotic formulas *in conjunction* with the elementary foundational rules of summability calculus.

The second building block of summability calculus is asymptotic analysis. Ever since Stirling and Euler presented their famous asymptotic expressions to the factorials and harmonic numbers respectively, a deep interest in the asymptotic behavior of special functions, such as finite sums, was forever instilled in the hearts of mathematicians. This includes Glaisher's work on the hyperfactorial function and Barnes work on the superfactorial function, to name a few. Here, the Gregory quadrature formula, the Euler-Maclaurin summation formula, and the Boole summation formula were proved to be indispensable. For example, a very nice application of such asymptotic formulas in explaining a curious observation of approximation errors is illustrated in [BBD89]. In this monograph, we will show how such formulas can be deduced from basic principles and how they can be used to perform infinitesimal calculus despite their usual divergence. We will also prove the equivalence of all asymptotic formulas.

The third building block of summability calculus is summability theory. Perhaps the most famous summability theorist was again Euler, who did not hesitate to use divergent series. According to Euler, the sum of an infinite series should carry a more general definition. In particular, if a sequence of algebraic operations eventually arrives at a divergent series, the value of the divergent series should assume the value of the algebraic expression from which it was derived [Kli83]. For example, since the series $1 - x + x^2 - x^3 + x^4 \ldots$ is formally equivalent to the function $f(x) = (1 + x)^{-1}$, one can assign the value of $\frac{1}{2}$ to the Grandi series. However, Euler's approach was ad-hoc based, and the first systematic and coherent theory of divergent series was initiated by Cesàro, who introduced what is now referred to as the Cesàro mean [Har49]. The Cesàro summability method, although weak, enjoys the advantage of being stable, regular, and linear, which will prove to be crucial for mathematical consistency. In the twentieth century, the most well-renowned summability theorist was Hardy, whose classic book *Divergent Series* remains authoritative.

Earlier, we stated that summable divergent series behaved "as if they were convergent". So, in principle, we would expect a finite sum to satisfy Eq. 1.3.2 if the infinite sums are summable. Ramanujan realized that the summability of divergent series might be linked to infinitesimal calculus using this equation [Ber85]. However, his statements were often imprecisely stated, and his conclusions were occasionally incorrect for reasons that will become clear later. In this monograph, we state results precisely and prove correctness. We will present the generalized definition of the value of a series $\mathfrak{T}$ in Chap. 4, which allows us to integrate the subject of summability theory with the study of finite sums into a single coherent framework. As will be shown repeatedly throughout this monograph, this generalized definition of the value of series is indeed one of the cornerstones of summability calculus. It is close in spirit to Euler's ideas, but with some slight additional restrictions. It will also shed important insights into the subject of divergent series. For example, we will present a method of deducing analytic expressions to, and "accelerating convergence" of, divergent series. In addition, we will also introduce a new summability method χ, which is weaker than $\mathfrak{T}$ but it is strong enough to "correctly" evaluate almost all of the examples of divergent series mentioned in this monograph. So, whereas $\mathfrak{T}$ is a formal definition that is decoupled from any particular method of computation, results will be deduced using $\mathfrak{T}$ and verified numerically using χ. The meaning of such statements will become clear in Chap. 4.

The fourth building block of summability calculus is polynomial approximation. Surprisingly, little work has been made to define fractional finite sums using polynomial approximations, except notably for the fairly recent work of Müller and Schleicher [MS10]. In their work, a fractional finite sum is defined by examining its asymptotic behavior. In a nutshell, if a finite sum can be approximated *in a bounded region* by a polynomial, and if the error of such an approximation vanishes as this bounded region is pushed towards infinity, we might then evaluate a fractional sum by evaluating it asymptotically and propagating results backwards

using the recurrence identity in Eq. 1.3.1. This method is described and extended in
length in Chap. 6. In 2011, moreover, Müller and Schleicher provided an *axiomatic*
treatment to the subject of finite sums by proposing six axioms that uniquely
extend the definition of finite sums to non-integer arguments, two of which are the
defining properties of finite sums listed earlier [MS11]. In addition to those two
properties, they included linearity, holomorphicity of finite sums of polynomials,
and translation invariance (the reader is referred to [MS11] for details). Perhaps
most importantly, they also included the method of polynomial approximation as a
sixth axiom; meaning that if a simple finite sum $\sum_{k=0}^{n-1} g(k)$ can be approximated by
a polynomial asymptotically, then such an approximation is assumed to be valid for
fractional values of n.

We do not pursue an axiomatic foundation of finite sums, although the treatment
of Müller and Schleicher is a special case of a more general definition that is
presented in this monograph because their work is restricted to simple finite sums
$\sum_{k=0}^{n-1} g(k)$, in which $g(n)$ is of a finite polynomial order (see Definition 1.1). In this
monograph, on the other hand, we will present a general statement that is applicable
to all finite sums, including those that cannot be approximated by polynomials, and
extend it further to oscillating sums and composite sums as well. Finally, this is
described in details in Chap. 6.

Last but not the least, the fifth building block of summability calculus is the
calculus of finite differences, which was first systematically developed by Jacob
Stirling in 1730 [Jor65], although some of its most basic results can be traced back to
the works of Newton, such as the Newton interpolation formula. In this monograph,
we will derive the basic results of the calculus of finite differences from summability
calculus and show that the two fields are closely related to each other. For example,
we will show that the summability method χ introduced in Chap. 4 is intimately tied
to Newton's interpolation formula, and present a geometric proof to the sampling
theorem using the generalized definition $\mathfrak{T}$ and the calculus of finite differences. In
fact, we will be able to prove a slightly stronger statement. The sampling theorem
was popularized in the twentieth century by Claude Shannon in his seminal paper
"Communications in the Presence of Noise" [Sha49], but its origin dates much
earlier. An excellent brief introduction to the sampling theorem and its origins is
given in [Luk99].

Summability calculus yields a rich set of identities related to fundamental
constants, such as γ, π, and e, and special functions, such as the Riemann zeta
function $\zeta(s)$ and the gamma function $\Gamma(n)$. Some of those identities shown in this
monograph appear to be new while others are known to be old. It is intriguing to
note that while many of these identities were proved at different periods of time by
different mathematicians using different approaches, summability calculus provides
simple tools for deriving all of them in an elementary manner.

1.6   Outline of Work

The rest of this monograph is structured as follows. We will first introduce the foundational rules of performing infinitesimal calculus, such as differentiation, integration, and computing series expansion, on simple finite sums in Chap. 2, and extend the calculus next to address composite finite sums in Chap. 3. After that, we introduce the generalized definition $\mathfrak{T}$ for the value of a series in Chap. 4 and show how central it is to summability calculus. Using $\mathfrak{T}$, we derive in Chap. 5 some foundational theorems for oscillating finite sums that simplify their study considerably. These theorems will yield important insights into the study of divergent series and series convergence acceleration. Next, we present a simple method for directly *evaluating* fractional finite sums $\sum_{k=0}^{n-1} s_k \, g(k, n)$ for all $n \in \mathbb{C}$ in Chap. 6. Using these results, we finally extend the calculus to arbitrary discrete functions, which leads immediately to some of the most fundamental results in the calculus of finite differences in Chap. 7. Throughout these sections, we will repeatedly prove that we are always dealing with the same generalized definition of finite sums, meaning that all results are consistent with each other and can be used interchangeably.

References

[Ala12] I. Alabdulmohsin, *Summability Calculus* (2012). arXiv:1209.5739v1 [math.CA]

[Apo99] T.M. Apostol, An elementary view of Euler's summation formula. Am. Math. Mon. **106**(5), 409–418 (1999)

[Ber85] B.C. Berndt, *Ramanujan's Notebooks*. Ramanujan's Theory of Divergent Series (Springer, New York, 1985)

[BS75] B. Berndt, L. Schoenfeld, Periodic analogues of the Euler-Maclaurin and Poisson summation formulas with applications to number theory. Acta Arith. **28**(1), 23–68 (1975)

[BBD89] J.M. Borwein, P.B. Borwein, K. Dilcher, Pi, Euler numbers, and asymptotic expansions. Am. Math. Mon. **96**(8), 681–687 (1989)

[BCM09] J.M. Borwein, N.J. Calkin, D. Manna, Euler-Boole summation revisited. Am. Math. Mon. **116**(5), 387–412 (2009)

[CS97] J. Choi, H.M. Srivastava, Sums associated with the zeta function. J. Math. Anal. Appl. **206**(1), 103–120 (1997)

[CS00] J. Choi, H.M. Srivastava, Certain classes of series associated with the zeta function and multiple gamma functions. J. Comput. Appl. Math. **118**, 87–109 (2000)

[Dav59] P.J. Davis, Leonhard Euler's integral: a historical profile of the gamma function. Am. Math. Mon. **66**(10), 849–869 (1959)

[Eul38a] L. Euler, General methods of summing progressions. Commentarii academiae scientiarum imperialis Petropolitanae **6**, 68–97 (1738). English Translation by I. Bruce at: http://www.17centurymaths.com/contents/euler/e025tr.pdf

[Eul38b] L. Euler, On transcendental progressions. that is, those whose general terms cannot be given algebraically. Commentarii academiae scientiarum imperialis Petropolitanae **5**, 36–57 (1738). English Translation by S. Langton at: http://home.sandiego.edu/~langton/eg.pdf

[Gla93] J. Glaisher, On certain numerical products in which the exponents depend upon the numbers. Messenger Math. **23**, 145–175 (1893)

[Har49] G.H. Hardy, *Divergent Series* (Oxford University Press, New York, 1949)

[Jor65] K. Jordán, *Calculus of Finite Differences* (Chelsea Publishing Company, New York, 1965)

[Kli83] M. Kline, Euler and infinite series. Math. Mag. **56**(5), 307–314 (1983)

[Kra99] S.G. Krantz, The Bohr-Mollerup theorem, in *Handbook of Complex Variables* (Springer, Berlin, 1999), p. 157

[Lam01] V. Lampret, The Euler-Maclaurin and Taylor formulas: twin, elementary derivations. Math. Mag. **74**(2), 109–122 (2001)

[Lan64] C. Lanczos, A precision approximation of the gamma function. J. Soc. Ind. Appl. Math., Ser. B: Numer. Anal. **1**, 86–96 (1964)

[LRSC01] C. Leiserson, R. Rivest, C. Stein, T.H. Cormen, *Introduction to Algorithms* (The MIT Press, Cambridge, 2001)

[Luk99] H.D. Luke, The origins of the sampling theorem. IEEE Commun. Mag. **37**, 106–108 (1999)

[MS10] M. Müller, D. Schleicher, Fractional sums and Euler-like identities. Ramanujan J. **21**(2), 123–143 (2010)

[MS11] M. Müller, D. Schleicher, How to add a noninteger number of terms: from axioms to new identities. Am. Math. Mon. **118**(2), 136–152 (2011)

[Rob55] H. Robbins, A remark on Stirling's formula. Am. Math. Mon. **62**, 26–29 (1955)

[San03] E. Sandifer, How Euler did it: estimating the Basel problem, December 2003. Available at The Mathematical Association of America (MAA). http://www.maa.org/news/howeulerdidit.html

[San07] E. Sandifer, How Euler did it: gamma the constant, October 2007. Available at The Mathematical Association of America (MAA). http://www.maa.org/news/howeulerdidit.html

[Sha49] C.E. Shannon, Communication in the presence of noise. Proc. IEEE **37**, 10–21 (1949)

[SP11] S. Sinha, S. Pradhan, *Numerical Analysis & Statistical Methods* (Academic Publishers, Kolkata, 2011)

[Spo94] J.L. Spouge, Computation of the gamma, digamma, and trigamma functions. SIAM J. Numer. Anal. **31**(3), 931–944 (1994)

[Var07] V.S. Varadarajan, Euler and his work on infinite series. Bull. Am. Math. Soc. **44**, 515–539 (2007)

[Vol77] A. Volodarsky, Mathematical achievements of Aryabhata. Indian J. Hist. Sci. **12**(2), 147–149 (1977)

[Weia] E. Weisstein, Barnes G-function. http://mathworld.wolfram.com/BarnesG-Function.html

[Weib] E. Weisstein, Gamma function. http://mathworld.wolfram.com/GammaFunction.html

[Weic] E. Weisstein, Glaisher-Kinkelin constant. http://mathworld.wolfram.com/Glaisher-KinkelinConstant.html

[Weid] E. Weisstein, Hyperfactorial. http://mathworld.wolfram.com/Hyperfactorial.html

Chapter 2
Simple Finite Sums

Simplicity is the ultimate sophistication
Leonardo da Vinci (1452–1519)

Abstract We will begin our treatment of summability calculus by analyzing what will be referred to, throughout this book, as simple finite sums. Even though the results of this chapter are particular cases of the more general results presented in later chapters, they are important to start with for a few reasons. First, this chapter serves as an excellent introduction to what summability calculus can markedly accomplish. Second, simple finite sums are encountered more often and, hence, they deserve special treatment. Third, the results presented in this chapter for simple finite sums will, themselves, be used as building blocks for deriving the most general results in subsequent chapters. Among others, we establish that fractional finite sums are well-defined mathematical objects and show how various identities related to the Euler constant as well as the Riemann zeta function can actually be derived in an elementary manner using fractional finite sums.

We will begin our treatment of summability calculus by analyzing *simple* finite sums, which are the finite sums of the form $\sum_{k=0}^{n-1} g(k)$. Even though the results of this chapter are particular cases of the more general results presented in later chapters, they are important to start with for a few reasons. First, this chapter serves as an excellent introduction to what summability calculus can markedly accomplish. Second, simple finite sums are encountered more often and, hence, they deserve special treatment. Third, the results presented in this chapter for simple finite sums will, themselves, be used as building blocks for establishing the most general results in subsequent chapters.

2.1   Foundations

Suppose we have the simple finite sum $f(n) = \sum_{k=0}^{n-1} g(k)$ for some analytic function g. Our goal is to obtain a "natural" method of extending the domain of $f(n)$ to the complex plane $\mathbb{C}$. Let us denote this new function by $f_G(n) : \mathbb{C} \to \mathbb{C}$, not to be confused with its discrete counterpart $f(n)$. As shown in Sect. 1.3, the two defining properties of finite sums imply that the following recurrence identity holds for all $n \in \mathbb{C}$ if $f_G(n)$ exists:

$$f_G(n) = g(n-1) + f_G(n-1) \qquad (2.1.1)$$

Because $f_G(n) : \mathbb{C} \to \mathbb{C}$ is defined in the complex plane $\mathbb{C}$ by assumption, it follows by linearity of differentiation that the following identity must hold:

$$f'_G(n) = g'(n-1) + f'_G(n-1) \qquad (2.1.2)$$

An alternative proof of Eq. 2.1.2 can be made using the two defining properties of finite sums as follows:

$$f'_G(n) - f'_G(n-1) = \lim_{h \to 0} \frac{1}{h} \{ \sum_{k=0}^{n-1+h} g(k) - \sum_{k=0}^{n-1} g(k) - \sum_{k=0}^{n-2+h} g(k) + \sum_{k=0}^{n-2} g(k) \}$$

$$= \lim_{h \to 0} \frac{1}{h} \{ \sum_{k=0}^{n-1+h} g(k) - \sum_{k=0}^{n-2+h} g(k) - \sum_{k=0}^{n-1} g(k) + \sum_{k=0}^{n-2} g(k) \}$$

$$= \lim_{h \to 0} \frac{1}{h} \{ \sum_{k=n-1+h}^{n-1+h} g(k) - \sum_{k=n-1}^{n-1} g(k) \} \qquad \text{(by property 2)}$$

$$= \lim_{h \to 0} \frac{1}{h} \{ g(n-1+h) - g(n-1) \} \qquad \text{(by property 1)}$$

$$= g'(n-1)$$

So, Eq. 2.1.2 indeed holds. Such a recurrence identity for $f'_G(n)$ suggests that $f'_G(n)$ is itself a simple finite sum, except, possibly, for an additional periodic function. Thus, we know that the general solution must be of the form:

$$f'_G(n) = \sum_{k=0}^{n-1} g'(k) + p_1(n), \qquad (2.1.3)$$

where $p_1(n)$ is a periodic function with a unit period; i.e. $p_1(n + 1) = p_1(n)$. Differentiating Eq. 2.1.2 many times yields the following set of conditions:

$$f_G^{(r)}(n) = \sum_{k=0}^{n-1} g^{(r)}(k) + p_r(n), \qquad (2.1.4)$$

where, again, $p_r(n)$ are periodic functions with unit periods.

To determine the (unique) form of the periodic functions $p_r(n)$, we use a *successive polynomial approximation* method. We will show that this successive approximation method implies a *unique* definition of $f_G(n)$. In particular, the periodic functions $p_r(n)$ will turn out to be nothing but *constants*. However, these constants have *definite* values. Later in this chapter, we present methods for computing the values of those constants.

2.1.1 A Statement of Uniqueness

The successive polynomial approximation method works as follows. We have a simple finite sum $f(n) = \sum_{k=0}^{n-1} g(k)$ and we would like to extend its domain to the complex plane $\mathbb{C}$. Suppose that we restrict our attention, first, to the domain $[0, 2]$. We would like to a define a function $f_G(n)$ over the interval $[0, 2]$ such that $f_G(n)$ agrees with $f(n)$ for all $n \in \mathbb{N}$ and $f_G(n)$ satisfies the recurrence identity in Eq. 2.1.1 for all $n \in \mathbb{C}$.

From the recurrence identity in Eq. 2.1.1, we note that $f(0) = 0$. This, in fact, is a general result that holds for all simple finite sums. It is commonly referred to as the *empty sum*, and is often defined to be zero by convention. However, we note here that it must, in fact, be zero because of the recurrence identity in Eq. 2.1.1. Because this fact will be cited quite frequently in the sequel, we emphasize it in the following proposition:

Proposition 2.1 (Empty Sum) *In any simple finite sum* $f(n) = \sum_{k=0}^{n-1} g(k)$, *we have* $f(0) = 0$.

Now, in order to find a *continuous* function $f_G(n)$ defined in the interval $[0, 2]$, which satisfies the recurrence identity in Eq. 2.1.1 and agrees with $f(n)$ for all $n \in \mathbb{N}$, we can choose *any* function in the interval $[0, 1]$ such that $f_G(0) = 0$ and $f_G(1) = g(0)$. Then, we can define the function $f_G(n)$ in the interval $[1, 2]$ using the recurrence identity. Let us see what happens if we do this, and let us examine how the function changes as we try to make $f_G(n)$ smoother.

First, we define $f_G(n)$ as follows:

$$f_G(n) = \begin{cases} a_0 + a_1 n & \text{if } n \in [0, 1] \\ g(n-1) + f_G(n-1) & \text{if } n > 1 \end{cases} \tag{2.1.5}$$

The motivation behind choosing a linear function in the interval $[0, 1]$ is because any continuous function can be approximated arbitrarily well by polynomials (see Weierstrass Theorem [Ros80]). So, we will initially choose a linear function and add higher degrees later to make the function smoother. To satisfy the conditions $f_G(0) = 0$ and $f_G(1) = g(0)$, we must have:

$$f_G(n) = \begin{cases} g(0) n & \text{if } n \in [0, 1] \\ g(n-1) + f_G(n-1) & \text{if } n > 1 \end{cases} \tag{2.1.6}$$

Clearly, Eq. 2.1.6 is a continuous function that satisfies the required recurrence identity and boundary conditions. However, its derivative $f'_G(n)$ is, in general, discontinuous at $n = 1$. To improve our polynomial approximation so that both $f_G(n)$ and $f'_G(n)$ are continuous throughout the interval $[0, 2]$, we make $f_G(n)$ a polynomial of the second degree in the interval $[0, 1]$. This yields:

$$f_G(n) = \begin{cases} \left(\frac{g(0)}{1} - \frac{g'(0)}{2}\right)n + \frac{g'(0)}{2}n^2 & \text{if } n \in [0, 1] \\ g(n-1) + f_G(n-1) & \text{if } n > 1 \end{cases} \tag{2.1.7}$$

In Eq. 2.1.7, both $f_G(n)$ and $f'_G(n)$ are continuous throughout the interval $[0, 2]$, and the function $f_G(n)$ satisfies the recurrence and boundary conditions. However, the second derivative is now discontinuous. Again, to make it continuous, we improve our estimate by making $f_G(n)$ a polynomial of the third degree in the interval $[0, 1]$ and enforcing the appropriate conditions. This yields:

$$f_G(n) = \begin{cases} \left(\frac{g(0)}{1} - \frac{g'(0)}{2} + \frac{g^{(2)}(0)}{12}\right)\frac{n}{1!} + \left(\frac{g'(0)}{1} - \frac{g^{(2)}(0)}{2}\right)\frac{n^2}{2!} + \frac{g^{(2)}(0)}{1}\frac{n^3}{3!} & \text{if } n \in [0, 1] \\ g(n-1) + f_G(n-1) & \text{if } n > 1 \end{cases} \tag{2.1.8}$$

Now, we begin to see a curious trend. It appears that in order for the function to satisfy the recurrence identity and boundary conditions and at the same time be infinitely differentiable, the successive polynomial approximation method implies that the rth derivative is formally given by $f_G^{(r)}(0) = \sum_{k=0}^{\infty} \frac{B_k}{k!} g^{(k+r-1)}(0)$, where $B_k = \{1, -\frac{1}{2}, \frac{1}{12}, \ldots\}$. Indeed this result will be established next, where the constants B_k are the Bernoulli numbers!

Theorem 2.1 *Given a simple finite sum* $f(n) = \sum_{k=0}^{n-1} g(k)$, *where* $g(k)$ *is analytic at the origin, let* $p_r(n)$ *to be a polynomial of degree* r, *and define* $f_{G,r}(n)$ *by:*

$$f_{G,r}(n) = \begin{cases} p_r(n) & \text{if } n \in [0, 1] \\ g(n-1) + f_{G,r}(n-1) & \text{otherwise} \end{cases}$$

If we require that $f_{G,r}(n)$ *be* $(r-1)$-*times differentiable in the domain* $[0, 2]$, *then the sequence of polynomials* $p_r(n)$ *is unique. In particular, its limit* $f_G(n) = \lim_{r \to \infty} f_{G,r}(n)$ *is unique and satisfies both the initial condition* $f_G(0) = 0$ *and the recurrence identity* $f_G(n) = g(n-1) + f_G(n-1)$.

Proof Writing:

$$f_{G,r}(n) = \begin{cases} \sum_{k=1}^{r} \frac{a_k}{k!} n^k & \text{if } n \in [0, 1] \\ g(n-1) + f_{G,r}(n-1) & \text{if } n > 1 \end{cases} \tag{2.1.9}$$

We would like to show that the boundary condition $f_G(0) = 0$, the recurrence identity $f_G(n) = g(n-1) + f_G(n-1)$, and the requirement of having $f_{G,r}(n)$ be

$(r-1)$-times differentiable translate into a non-singular system of linear equations on a_k, whose solution must, therefore, be unique.

The conditions of continuity in $f_G(n)$ and its first $(r-1)$ derivatives imply that:

$$
\begin{aligned}
\frac{a_1}{1!} + \frac{a_2}{2!} + \frac{a_3}{3!} + \cdots + \cdots + \frac{a_r}{r!} &= g(0) \\
\frac{a_2}{1!} + \frac{a_3}{2!} + \cdots + \cdots + \frac{a_r}{(r-1)!} &= g'(0) \\
\frac{a_3}{1!} + \cdots + \cdots + \frac{a_r}{(r-2)!} &= g''(0) \\
\cdots \quad \cdots \quad \cdots & \\
\frac{a_r}{0!} &= g^{(r)}(0)
\end{aligned}
$$

Since the system is non-singular, the values of $a_1, \ldots, a_r$ are unique. Hence, the polynomials $p_r(n) = \sum_{k=1}^{n}(a_k/k!)n^k$ are unique as well. Using elementary properties of the Bernoulli numbers, we deduce that the solution is given by:

$$
a_k = \sum_{j=0}^{r-k} \frac{B_j}{j!} g^{(j+k-1)}(0) \qquad \square
$$

Next, we use Theorem 2.1 to deduce a matrix identity related to the Bernoulli numbers, which we can use as a definition of those important numbers.

Corollary 2.1 *Let B_r be the Bernoulli numbers $(1, -\frac{1}{2}, \frac{1}{6}, 0, \ldots)$. Then:*

$$
\begin{bmatrix}
\frac{B_0}{0!} & \frac{B_1}{1!} & \frac{B_2}{2!} & \cdots\cdots & \frac{B_{r-1}}{(r-1)!} \\
0 & \frac{B_0}{0!} & \frac{B_1}{1!} & \cdots\cdots & \frac{B_{r-2}}{(r-2)!} \\
0 & 0 & \frac{B_0}{0!} & \cdots\cdots & \frac{B_{r-3}}{(r-3)!} \\
\cdots & \cdots & \cdots & \cdots\cdots & \cdots \\
\cdots & \cdots & \cdots & \cdots\cdots & \cdots \\
0 & 0 & 0 & \cdots\cdots & \frac{B_0}{0!}
\end{bmatrix}
=
\begin{bmatrix}
\frac{1}{1!} & \frac{1}{2!} & \frac{1}{3!} & \cdots\cdots & \frac{1}{r!} \\
0 & \frac{1}{1!} & \frac{1}{2!} & \frac{1}{3!} & \cdots\cdots & \frac{1}{(r-1)!} \\
0 & 0 & \frac{1}{1!} & \frac{1}{2!} & \frac{1}{3!} & \cdots & \frac{1}{(r-2)!} \\
\cdots & \cdots & \cdots & \cdots\cdots & \cdots \\
\cdots & \cdots & \cdots & \cdots\cdots & \cdots \\
\cdots & \cdots & \cdots & \cdots\cdots & \frac{1}{1!}
\end{bmatrix}^{-1}
$$

Proof This follows from Theorem 2.1. $\qquad\square$

2.1.2 Differentiation and Integration Rules

Now, we return to our original question. What are the periodic functions $p_r(n)$ given in Eq. 2.1.4? If we use the successive polynomial approximation method as a *definition* for fractional finite sums, we obtain the following formal expression:

$$
f_G(n) = \sum_{k=1}^{\infty} \frac{n^k}{k!} \sum_{j=0}^{\infty} \frac{B_j}{j!} g^{(k+j-1)}(0) \tag{2.1.10}
$$

By formally differentiating with respect to n, we obtain:

$$f_G^{(r)}(n) = \sum_{k=1}^{\infty} \frac{n^k}{k!} \sum_{j=0}^{\infty} \frac{B_j}{j!} g^{(k+j-1+r)}(0) + \sum_{j=0}^{\infty} \frac{B_j}{j!} g^{(j+r)}(0) \qquad (2.1.11)$$

$$= \sum_{k=0}^{n-1} g^{(r)}(k) + c_r, \qquad (2.1.12)$$

where c_r is formally given by $\sum_{j=0}^{\infty} \frac{B_j}{j!} g^{(j+r)}(0)$, which is independent of n. Hence, the periodic functions $p_r(n)$ are constants. These constants have definite values, and we will show how they are computed later.

Proposition 2.2 (Differentiation Rule) *The derivative of a simple finite sum* $f(n) = \sum_{k=0}^{n-1} g(k)$ *is given by:*

$$f'(n) = \sum_{k=0}^{n-1} g'(k) + c, \qquad (2.1.13)$$

for some definite, non-arbitrary constant $c = f'(0)$.

From the differentiation rule above, we obtain the corresponding integration rule:

Proposition 2.3 (Integration Rule) *The indefinite integral of a simple finite sum* $f(n) = \sum_{k=0}^{n-1} g(k)$ *is given by:*

$$\int^n \sum_{k=0}^{t-1} g(k)\,dt = \sum_{k=0}^{n-1} \int^k g(t)\,dt + c_1 n + c_2 \qquad (2.1.14)$$

for some definite, non-arbitrary constant $c_1 = -\frac{d}{dn} \sum_{k=0}^{n-1} \int^k g(t)\,dt \Big|_{n=0}$ *and some arbitrary constant* c_2.

These results form the preliminary building blocks for summability calculus.

2.1.3 Examples

The differentiation and integrations rules in Propositions 2.2 and 2.3 are incomplete; we still do not know how to compute the values of the non-arbitrary constants. We will present later a procedure for computing the values of those constants. Before then, we present two examples in this section, in which the non-arbitrary constants can be determined using some elementary methods.

2.1.3.1 The Power Sum Function

For our first example, we return to the power sum function presented earlier in Eq. 1.1.5. The first point we note about this function is that its $(m + 2)$th derivative is **zero**, which follows immediately from the differentiation rule in Proposition 2.2. Thus, the *unique* generalized definition of the power sum function that is dictated by Theorem 2.1 has to be a polynomial of degree $(m + 1)$. As mentioned earlier in Chap. 1, those polynomials are given by the Bernoulli-Faulhaber formula.

Second, writing $f(n; m) = \sum_{k=0}^{n-1} k^m$, we obtain by the integration rule in Proposition 2.3:

$$\int_0^n \sum_{k=0}^{t-1} k^m \, dt = \frac{1}{m+1} \sum_{k=0}^{n-1} k^{m+1} + c_1 n + c_2 \tag{2.1.15}$$

Using the empty sum rule, i.e. Proposition 2.1, we conclude that $c_2 = 0$. Setting $n = 1$ in both sides gives us the desired value for c_1:

$$c_1 = \int_0^1 f_G(t; m) dt$$

This gives us the following simple recursive rule for deducing closed-form expressions of power sums:

$$f_G(n; m + 1) = (m + 1) \int_0^n f_G(t; m) \, dt - (m + 1) \left(\int_0^1 f_G(t; m) \, dt \right) n \tag{2.1.16}$$

For example, knowing that triangular numbers satisfy:

$$f_G(n; 1) = \sum_{k=0}^{n-1} k = \frac{n(n-1)}{2}$$

We deduce from Eq. 2.1.16 that the following holds as well:

$$f_G(n; 2) = \sum_{k=0}^{n-1} k^2 = 2 \int_0^n \frac{t(t-1)}{2} \, dt - 2n \int_0^1 \frac{t(t-1)}{2} \, dt$$

$$= \frac{n^3}{3} - \frac{n^2}{2} + \frac{n}{6}$$

Third, because we always have $f_G(0; m) = 0$ and $f_G(1; m) = 0$ for $m \geq 1$, we conclude that $n(n-1)$ is always a proper factor of power sum polynomials. The fact that $n(n-1)$ is always a proper factor was observed as early as Faulhaber himself in his book *Academia Algebrae* in 1631 [Hav03]. The recursive solution of power sums given in Eq. 2.1.16 is well known and was used by Jacob Bernoulli [Spi06, Blo93].

Due to its apparent simplicity, it has been called the "Cook-Book Recipe" by some mathematicians [Shi07]. Nevertheless, our simple three-line proof here illustrates the efficacy of summability calculus in analyzing finite sums.

2.1.3.2 A Second Example

Our second example is the function:

$$f(n) = \sum_{k=0}^{n-1} k x^k \tag{2.1.17}$$

Before we derive an analytic expression for the simple finite sum above using summability calculus, we know in advance that $f_G(0) = f_G(1) = 0$ by the empty sum rule and by plugging $n = 1$, respectively. This may serve as a sanity check for us later.

Using the differentiation rule for simple finite sums, we have:

$$f_G'(n) = \sum_{k=0}^{n-1} x^k + \log x \, f_G(n) + c = \frac{1 - x^n}{1 - x} + \log x \, f_G(n) + c \tag{2.1.18}$$

Equation 2.1.18 is a first-order linear differential equation whose solution is available in closed-form [TP85]. Using the initial condition $f_G(1) = 0$, and after rearranging the terms, we obtain the closed-form expression given Eq. 2.1.19. Note that $f_G(0) = f_G(1) = 0$ holds as expected.

$$f_G(n) = \frac{1}{(1 - x)^2} \left(x^n (nx - n - x) + x \right) \tag{2.1.19}$$

Exercise 2.1 Use the differentiation rule of Proposition 2.2 to prove that $\frac{d}{dn} x^n = c x^n$ for some constant c that is independent of n. Observe again why the constant of differentiation in Proposition 2.2 has a definite value. (Hint: use the fact that $\log x^n = \sum_{k=0}^{n-1} \log x$)

Exercise 2.2 Show that the derivative of the log factorial function $f(n) = \log n!$ is given by $H_n - \gamma$, where H_n is the harmonic number and γ is some definite constant. Later, we will provide the machinery to prove that γ is, in fact, the Euler constant $0.5772166 \cdots$.

Exercise 2.3 Consider the function $f(n) = \sum_{k=0}^{n-1} \sin k$. Use the differentiation rule of Proposition 2.2 and the empty sum rule of Proposition 2.1 to prove that: $f(n) = \alpha \sin n + \beta (1 - \cos n)$, for some constants α and β. Can we determine the values of α and β using the definition of $f(n)$? (Hint: you may need to differentiate many times and collect all terms in a Taylor series expansion)

2.2 Semi-linear Simple Finite Sums

Next, we begin to unravel the general formulas for the definite constants of the differentiation and integration rules. We will begin with the simplest case, which we refer to as *semi-linear* simple finite sums. These are finite sums of the form $\sum_{k=0}^{n-1} g(k)$, where g is of a *polynomial order* zero and satisfies a few additional conditions (see Definition 1.1). The results we present here will be extended in the next section to the more general case of arbitrary simple finite sums.

We start with a few preliminary definitions.

Definition 2.1 (Nearly-Convergent Functions) A function $g : \mathbb{C} \to \mathbb{C}$ is called nearly convergent if $\lim_{k \in \mathbb{R}: \, k \to \infty} g'(k) = 0$ (i.e. it is of a polynomial order zero) and one of the following two conditions holds:

1. g is asymptotically non-decreasing and concave. More precisely, there exists $k_0 \in \mathbb{R}$ such that for all $k > k_0$, we have $g'(k) \geq 0$ and $g^{(2)}(k) \leq 0$.
2. g is asymptotically non-increasing and convex. More precisely, there exists $k_0 \in \mathbb{R}$ such that for all $k > k_0$, we have $g'(k) \leq 0$ and $g^{(2)}(k) \geq 0$

Definition 2.2 (Semi-linear Simple Finite Sums) A simple finite sum $f(n) = \sum_{k=0}^{n-1} g(k)$ is called semi-linear if g is a nearly-convergent function.

Informally speaking, a function g is nearly convergent if it is both asymptotically monotone and its rate of change vanishes asymptotically. In other words, $g(k)$ *becomes almost constant* in the bounded region $k \in (k_0 - W, k_0 + W)$ for any fixed $W \in \mathbb{R}$ as $k_0 \to \infty$. Semi-linear simple finite sums are quite common, e.g. $\sum_{k=0}^{n-1} k^m$ for $m < 1$ and $\sum_{k=0}^{n-1} \log^s (1+k)$, and summability calculus is quite simple for this important class of functions. Intuitively, because $g(k)$ is almost constant asymptotically in any bounded region, we expect $f'_G(n)$ to be close to $g(n)$ as $n \to \infty$. This is indeed the case as will be established soon. Nearly-convergent functions are depicted in Fig. 2.1.

Lemma 2.1 *If $g(k)$ is nearly convergent, then the limit* $\displaystyle \lim_{n \to \infty} \left\{ g(n-1) - \sum_{k=0}^{n-1} g'(k) \right\}$

exists.

Proof We will prove the lemma here for the case where $g(k)$ is asymptotically non-decreasing and concave. Similar steps can be used in the second case where $g(k)$ is asymptotically non-increasing and convex. First, let E_n and D_n be given by Eq. 2.2.1,

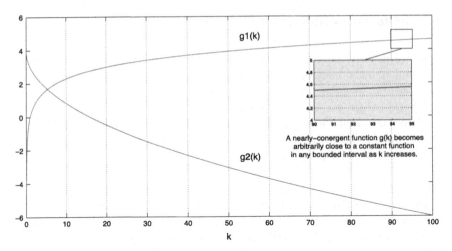

Fig. 2.1 Examples of nearly-convergent functions are depicted in this figure. The function $g_1(k)$ is asymptotically non-decreasing and concave, while $g_2(k)$ is asymptotically non-increasing and convex. In both cases, the derivatives also vanish as $k \to \infty$

where k_0 is defined as in Definition 2.1.

$$E_n = \sum_{k=k_0}^{n-1} g'(k) - g(n-1), \qquad D_n = E_n - g'(n-1) \qquad (2.2.1)$$

We will now show that the limit $\lim_{n\to\infty} E_n$ exists. Clearly, since $k_0 < \infty$, Lemma 2.1 follows immediately after words. By definition of E_n and D_n, we have:

$$E_{n+1} - E_n = g'(n) - \big(g(n) - g(n-1)\big) \qquad (2.2.2)$$

$$D_{n+1} - D_n = g'(n-1) - \big(g(n) - g(n-1)\big) \qquad (2.2.3)$$

Because $g'(n) \geq 0$ by assumption, $D_n \leq E_n$. Thus, D_n is a lower bound on E_n. However, concavity of $g(n)$ implies that $g'(n) \leq g(n) - g(n-1)$ and $g'(n-1) \geq g(n) - g(n-1)$. Placing these inequalities into Eqs. 2.2.2 and 2.2.3 implies that E_n is a non-increasing sequence while D_n is a non-decreasing sequence. Since D_n is a lower bound on E_n, E_n converges, which completes the proof of the lemma. □

Lemma 2.1 immediately proves not only the existence of Euler's constant but also *Stieltjes constants* and *Euler's generalized constants* as well (for a definition of these families of constants, the reader may refer to [Weid, Hav03]). Notice that so far, we have not made use of the fact that $\lim_{k\to\infty} g'(k) = 0$, but this last property will be important for our next main theorem.

Theorem 2.2 *Suppose* $g : \mathbb{C} \to \mathbb{C}$ *is regular at the origin. Let a simple finite sum be given by* $f(n) = \sum_{k=0}^{n-1} g(k)$ *, where* $g^{(m)}(k)$ *that denotes the* mth *derivative of* $g(k)$ *is a nearly-convergent function for all* $m \geq 0$. *Also, let* $f_G(n)$ *be given formally by the following Maclaurin series expansion:*

$$f_G(n) = \sum_{r=1}^{\infty} \frac{c_r}{r!} n^r, \qquad \text{where } c_r = \lim_{n \to \infty} \left\{ g^{(r-1)}(n-1) - \sum_{k=0}^{n-1} g^{(r)}(k) \right\} \qquad (2.2.4)$$

Then, $f_G(n)$ *satisfies formally the recurrence identity and the initial conditions given in Eq. 2.1.1.*

Proof Writing formally:

$$f_G(n) = \sum_{r=1}^{\infty} \frac{c_r}{r!} n^r \qquad (2.2.5)$$

We have:

$$f_G(n+1) - f_G(n) = \sum_{r=1}^{\infty} \frac{c_r}{r!} (n+1)^r - \sum_{r=1}^{\infty} \frac{c_r}{r!} n^r = \sum_{r=0}^{\infty} \frac{K_r}{r!} n^r,$$

where $K_r = \sum_{j=1}^{\infty} \frac{c_{j+r}}{j!}$. However, in order for the recurrence identity to hold, we must have:

$$K_r = g^{(r)}(0), \qquad \text{for all } r \geq 0 \qquad (2.2.6)$$

Luckily, the conditions in Eq. 2.2.6 for $r \geq 0$ are all *equivalent* by definition of c_r so we only have to prove that they hold for $r = 0$. This is because the condition in Eq. 2.2.6 for $r = 1$ is equivalent to the condition for $r = 0$ if we replace $g(k)$ with its derivative $g'(k)$, and so on. To prove that it holds for $r = 0$, we note that:

$$K_0 = \sum_{j=1}^{\infty} \frac{c_j}{j!} = \sum_{j=1}^{\infty} \frac{1}{j!} \lim_{n \to \infty} \left\{ g^{(j-1)}(n) - \sum_{k=0}^{n} g^{(j)}(k) \right\}$$

$$= \lim_{n \to \infty} \left\{ \sum_{j=1}^{\infty} \frac{g^{(j-1)}(n)}{j!} - \sum_{j=1}^{\infty} \frac{1}{j!} \sum_{k=0}^{n} g^{(j)}(k) \right\}$$

Using the formal expression:

$$\sum_{j=1}^{\infty} \frac{1}{j!} g^{(j-1)}(n) = \int_{n}^{n+1} g(t)\, dt \qquad (2.2.7)$$

Similarly, we have:

$$\sum_{j=1}^{\infty} \frac{1}{j!} g^{(j)}(s) = g(s+1) - g(s) \qquad (2.2.8)$$

From Eq. 2.2.8, we realize that:

$$\sum_{j=1}^{\infty} \frac{1}{j!} \sum_{k=0}^{n} g^{(j)}(k) = g(n+1) - g(0) \qquad (2.2.9)$$

This is because the left-hand sum is a telescoping sum. Plugging both Eqs. 2.2.7 and 2.2.9 into the last expression for K_0 yields:

$$K_0 = g(0) - \lim_{n \to \infty} \left\{ g(n+1) - \int_n^{n+1} g(t)\, dt \right\} \qquad (2.2.10)$$

Now, we need to show that the second term on the right-hand side evaluates to zero. This is easily shown upon noting that the function g is nearly-convergent. As stated earlier, since g is nearly convergent, then $g^{(m)}(n)$ vanishes asymptotically as $n \to \infty$ for all $m \geq 1$. Since g is also asymptotically monotone by Definition 2.1, then for any $\epsilon > 0$, there exists a constant N large enough such that:

$$\left| \max_{x \in [n, n+\tau]} g(x) - \min_{x \in [n, n+\tau]} g(x) \right| < \epsilon, \quad \text{for all } n > N \text{ and } \tau > 0 \qquad (2.2.11)$$

Consequently, we can find a constant N large enough such that:

$$\left| g(n+1) - \int_n^{n+1} g(t)\, dt \right| < \epsilon, \quad \text{for all } n > N \qquad (2.2.12)$$

Because ϵ can be made arbitrarily close to zero for which N has to be sufficiently large, we must have:

$$\lim_{n \to \infty} \left\{ g(n+1) - \int_n^{n+1} g(t)\, dt \right\} = 0 \qquad (2.2.13)$$

Therefore, we indeed have $K_0 = g(0)$. As established earlier, this implies that the conditions in Eq. 2.2.6 also hold.

With this in mind, we have established formally that $f_G(n+1) - f_G(n) = g(n)$, where $f_G(n)$ is given by Eq. 2.2.4. Hence, the recurrence identity is satisfied. Finally, proving that the initial condition also holds follows immediately from the empty sum run and the fact that $f_G(0) = 0$, which completes the proof of the theorem. □

Theorem 2.2 provides us with a very convenient method for performing infinitesimal calculus, such as differentiation, integration, and computing series expansions. As will be illustrated in Sect. 2.3, it can even be used to derive asymptotic expansions, such as the Stirling approximation. Before we discuss its implications, let us briefly summarize its statement.

Earlier, we stated that any simple finite sum of the form $f(n) = \sum_{k=0}^{n-1} g(k)$ satisfies:

$$f'_G(n) = \sum_{k=0}^{n-1} g'(k) \ + c$$

The constant c has a definite value that is independent of n. So, to evaluate the derivative $f'_G(n)$ using the above equation, we need to determine the value of c. However, this constant can be determined once we know, at least, one value of the derivative $f'_G(n)$ for some $n \in \mathbb{N}$. For semi-linear simple finite sums, we can compute the value of c because we know that $f'_G(n)$ asymptotically approaches $g(n)$ as $n \to \infty$. This is all the information we need to compute c because this fact along with the differentiation rule of simple finite sums both imply that:

$$c = \lim_{n \to \infty} \left\{ f'_G(n) - \sum_{k=0}^{n-1} g'(k) \right\} = \lim_{n \to \infty} \left\{ g(n) - \sum_{k=0}^{n} g'(k) \right\}$$

To recall, we proved that the limit exists in Lemma 2.1. Hence, we can use the formula above to compute c and to compute $f'_G(n)$ as a result.

Remark 2.1 We defer the proof that Theorem 2.2 is consistent with the successive polynomial approximation method of Theorem 2.1 until Sect. 2.4.

In the previous proof of Theorem 2.2, we showed that if g is nearly convergent and given any $\epsilon > 0$, we can always find a constant N large enough such that Eq. 2.2.11 holds. Here, Eq. 2.2.11 implies that the finite sum $f(n) = \sum_{k=0}^{n-1} g(k)$ grows *almost linearly* around n for a sufficiently large $n \gg 0$ and that its derivative can be made arbitrarily close to $g(n)$, which, in turn, is nearly constant as $n \to \infty$. We, therefore, expect its natural generalization to exhibit the same behavior and to become approximately linear asymptotically. This is achieved if $f'_G(n)$ is made arbitrarily close to $g(n)$ when $n \to \infty$. However, such a condition is precisely what Theorem 2.2 states, and, by Lemma 2.1, the limit exists so the function $f_G(n)$ given by Theorem 2.2 is indeed the *unique* function that satisfies such a property of natural generalization.

Now, we conclude with one last corollary. Here, we will show that if the limit $\lim_{n \to \infty} \sum_{k=0}^{n-1} g(k)$ exists, then an additional statement of the uniqueness of natural generalization can be made, which was touched upon earlier in Chap. 1.

Corollary 2.2 *Let the simple finite sum* $f(n) = \sum_{k=0}^{n-1} g(k)$ *be semi-linear and suppose that the series* $\sum_{k=0}^{\infty} g(k)$ *converges absolutely. Also, let* $f_G(n)$ *be the unique natural generalization that is given by Theorem 2.2. Then, we have for all* $n \in \mathbb{C}$:

$$f_G(n) = \sum_{k=0}^{\infty} g(k) - \sum_{k=n}^{\infty} g(k) \qquad (2.2.14)$$

Proof The expression in Eq. 2.2.14 is what we would expect from a unique natural generalization to simple finite sums if $\lim_{n\to\infty} \sum_{k=0}^{n-1} g(k)$ exists. This was discussed earlier in Chap. 1. To prove that Eq. 2.2.14 holds, define:

$$\tilde{f}(n) = \sum_{k=0}^{\infty} g(k) - \sum_{k=n}^{\infty} g(k) \qquad (2.2.15)$$

We would like to show that $\tilde{f}(n)$ and the generalization $f_G(n)$ of Theorem 2.2 are identical. If we employ the two defining properties of finite sums, we deduce that:

$$\sum_{k=n}^{\infty} g(k) = g(n) + g(n+1) + g(n+2) + \cdots \qquad (2.2.16)$$

Since the series converges absolutely, we take the derivative with respect to n of $\tilde{f}(n)$ term-by-term in Eq. 2.2.15, which reveals that all higher derivatives of $\tilde{f}(n)$ agree with the ones given by Theorem 2.2. In particular, we have:

$$\frac{d^r}{dn^r} \tilde{f}(n) = -\frac{d^r}{dn^r} \sum_{k=n}^{\infty} g(k)$$

$$= -\frac{d^r}{dn^r} \big(g(n) + g(n+1) + \cdots\big) = -\sum_{k=n}^{\infty} g'(k)$$

$$= \sum_{k=0}^{n-1} g^{(r)}(k) - \sum_{k=0}^{\infty} g^{(r)}(k)$$

On the other hand, Theorem 2.2 dictates that the unique most natural generalization $f_G(n)$ must satisfy for all $r \geq 1$:

$$\frac{d^r}{dn^r} f_G(n) = \sum_{k=0}^{n-1} g^{(r)}(k) + \lim_{n\to\infty} \Big\{ g^{(r-1)}(n) - \sum_{k=0}^{n} g^{(r)}(k) \Big\}$$

$$= \sum_{k=0}^{n-1} g^{(r)}(k) - \sum_{k=0}^{\infty} g^{(r)}(k)$$

Here, the last equality holds because $\lim_{n\to\infty} g^{(r-1)}(n) = 0$ under the stated conditions. Since both functions share the same higher order derivatives $\frac{d^r}{dn^r}f(n)$ for $r \geq 0$, they have the same Taylor series expansion. By the *uniqueness* of Taylor series expansions for analytic functions, i.e. rigidity, the two functions must be identical.   □

2.3 Examples to Semi-linear Simple Finite Sums

In this section, we present examples that illustrate the efficacy of the simple rules of summability calculus that have been developed so far in handling simple finite sums.

2.3.1 The Factorial Function

Let us consider the log-factorial function given by $\varpi(n) = \sum_{k=0}^{n-1} \log(1 + k)$. We can use summability calculus to derive the series expansion of $\varpi(n)$ quite easily as follows. First, by a direct application of Theorem 2.2, we find that the derivative at $n = 0$ is given by $\lim_{n\to\infty}\{\log n - H_n\} = -\gamma$, where H_n is the nth harmonic number and $\gamma = 0.5772\cdots$ is Euler's constant. Thus, we have by Rule 1:

$$\varpi'(n) = -\gamma + \sum_{k=0}^{n-1} \frac{1}{1+k} \tag{2.3.1}$$

Because $1/(1 + k)$ is also nearly convergent by Definition 2.1 and because it converges to zero, we have by Theorem 2.2:

$$\varpi^{(2)}(n) = \zeta(2) - \sum_{k=0}^{n-1} \frac{1}{(1+k)^2} \tag{2.3.2}$$

Here, $\zeta(s) = \sum_{k=1}^{\infty} \frac{1}{k^s}$ is the Riemann zeta function. Continuing in this manner yields the following series expansion:

$$\varpi(n) = \log n! = -\gamma n + \sum_{k=2}^{\infty} (-1)^k \frac{\zeta(k)}{k} n^k \tag{2.3.3}$$

In particular, since $\log 1 = 0$, we have the following well-known identity that was first proved by Euler [Seb02]:

$$\gamma = \sum_{k=2}^{\infty} (-1)^k \frac{\zeta(k)}{k} \tag{2.3.4}$$

In this example, it is important to keep a few points in mind. First, the series expansion in Eq. 2.3.3 is the series expansion of the unique most natural generalization of the log-factorial function, which turned out to be the series expansion of $\log \Gamma(n + 1)$, where Γ is the gamma function. However, our a priori belief on the nature of the generalized definition $f_G(n)$ may or may not be equivalent to what Theorem 2.2 implies. In other words, it was possible, at least in principle, that Theorem 2.2 would yield a generalized definition of the log-factorial function that is not related to the gamma function so one needs to exercise caution. In this particular example, nevertheless, they both turned out to be identical.

Second, using the earlier proofs, we deduce from this example that the function $\log \Gamma(n + 1)$ is the unique smooth function that satisfies the recurrence identity $f(n) = \log n + f(n - 1)$ and the initial condition $f(0) = 0$, and *its higher order derivatives do not alternate in sign infinitely many times*. Proposition 2.4 presents a more precise statement of the latter conclusion.

Proposition 2.4 *Let $f_G(n) : \mathbb{C} \to \mathbb{C}$ be a function that satisfies the following three properties:*

1. *$f_G(n) = \log n + f_G(n - 1)$, for all n*
2. *$f_G(0) = 0$*
3. *For every higher order derivative $f_G^{(r)}(n)$ where $r \geq 0$, there exists a constant n_r such that $f_G^{(r)}(n)$ is monotone for all $n > n_r$. In other words, $f_G^{(r+1)}(n)$ does not alternate in sign infinitely many times.*

Then, $f_G(n) = \log \Gamma(n + 1)$, where Γ is the gamma function.

Proof Let $f_G(n)$ be a function that satisfies $f_G(n) = \log n + f_G(n - 1)$ for all $n \in \mathbb{C}$ and it satisfies $f_G(0) = 0$. Because $\log(1 + n)$ is a nearly convergent function and because $f_G(n)$ is monotone for $n > n_0$, its derivative $f_G'(n)$ converges to $\log n$ as $n \to \infty$. However, this along with the differentiation rule of simple finite sums imply that $f_G'(0)$ is uniquely determined; its value must be equal to $-\gamma$. Similarly, since $f_G'(n) = -\gamma + H_n$, where the harmonic number is semi-linear, and because $f_G'(n)$ is monotone for all $n > n_1$, $f_G^{(2)}(0)$ is also uniquely determined and is given by $\zeta(2)$, and so on. Thus, all higher order derivatives at $n = 0$ are uniquely determined and are given by Theorem 2.2. Therefore, the only possible series expansion of $f_G(n)$ is the one given by Eq. 2.3.3, which is the series expansion of $\log \Gamma(n + 1)$. □

Proposition 2.4 presents a statement, which is different from the *Bohr-Mollerup theorem* [Kra99], as to why the gamma function is the unique most natural generalization of the discrete factorial function. According to the lemma, the log-gamma function is the only possible generalization of the log-factorial function if we require that higher derivatives do not alternate in sign infinitely many times. However, by requiring such a smooth asymptotic behavior of the function, its behavior *for all n* is given by $\log \Gamma(n + 1)$.

Historically, there have been several attempts to extend the factorial function into an entire function to avoid gamma's singularities at negative integers. One well-known example is Hadamard's factorial function, which correctly interpolates the discrete factorial function. However, Hadamard's function does not satisfy the required recurrence identity for all complex values of n [Dav59]. In fact, it is straightforward to observe that a function that satisfies the recurrence identity $f_G(n) = nf_G(n-1)$ for all n has to have singularities at negative integers, since we must have $1! = 1 \times 0!$ and $0! = 0 \times (-1)!$ by the recurrence identity. Since, both $0! = 1$ and $0! = 0 \times (-1)!$ must hold, $(-1)!$ has to be undefined. Therefore, the singularities at negative integers are unavoidable.

2.3.2 Euler's Constant

Suppose that $f(n)$ is given by the following equation:

$$f(n) = \log(n+x)! - \log x! = \sum_{k=0}^{n-1} \log(1+k+x) \tag{2.3.5}$$

In Eq. 2.3.5, let x be a constant that is independent of n. Using the differentiation rule of simple finite sums:

$$\frac{d}{dn}f_G(n) = c + \sum_{k=0}^{n-1} \frac{1}{1+k+x} \tag{2.3.6}$$

Using Theorem 2.2, we have:

$$c = \lim_{n \to \infty} \left\{ \log(n+x) - \sum_{k=0}^{n-1} \frac{1}{1+k+x} \right\}$$

$$= \lim_{n \to \infty} \left\{ \log(n+x) - \sum_{k=0}^{n+x} \frac{1}{1+k} \right\} + \lim_{n \to \infty} \left\{ \sum_{k=0}^{n+x} \frac{1}{1+k} - \sum_{k=0}^{n-1} \frac{1}{1+k+x} \right\}$$

$$= \lim_{n \to \infty} \left\{ \log(n+x) - \sum_{k=0}^{n+x-1} \frac{1}{1+k} \right\} + \lim_{n \to \infty} \left\{ \sum_{k=0}^{n+x-1} \frac{1}{1+k} - \sum_{k=x}^{n+x-1} \frac{1}{1+k} \right\}$$

$$= -\gamma + \sum_{k=0}^{x-1} \frac{1}{1+k} = -\gamma + H_x = \psi(1+x),$$

where ψ is the digamma function. Now, we employ Theorem 2.2 to derive the entire Taylor series expansion of $f_G(n)$. We have:

$$\frac{d}{dn}f_G(n) = \psi(x+1) + \sum_{k=2}^{\infty}(-1)^k\big(\zeta(k) - H_{k,x}\big)n^{k-1}, \tag{2.3.7}$$

where $H_{m,n} = \sum_{k=0}^{n-1} 1/(1+k)^m$ as defined earlier in Chap. 1. Next, we integrate both sides with respect to n, which yields:

$$f_G(n) = \psi(x+1)n + \sum_{k=2}^{\infty} \frac{(-1)^k}{k}(\zeta(k) - H_{k,x})n^k \qquad (2.3.8)$$

Here, we used the fact that $f_G(0) = 0$ by the empty sum rule. Since we derived earlier in Eq. 2.3.1 the identity $\psi(x+1) = \varpi'(x) = -\gamma + H_x$, we plug it into the previous equation and set $n = 1$ to deduce that:

$$\log(1+x) = -\frac{\gamma - H_{1,x}}{1} + \frac{\zeta(2) - H_{2,x}}{2} - \frac{\zeta(3) - H_{3,x}}{3} \cdots \qquad (2.3.9)$$

Thus, by taking the limit as $x \to \infty$, we recover Euler's famous result again:

$$\lim_{x \to \infty} \left\{ H_x - \log(1+x) - \gamma \right\} = 0 \qquad (2.3.10)$$

Of course, we could alternatively arrive at the same result by a direct application of Lemma 2.1. In addition, if we set $x = 0$ in Eq. 2.3.9 and use the empty sum rule, we arrive at Eq. 2.3.4 again. However, Eq. 2.3.9 is obviously a more general statement. For instance, by setting $x = 1$, we arrive at the identity in Eq. 2.3.11. Additionally, by plugging Eq. 2.3.4 into Eq. 2.3.9, we arrive at Eq. 2.3.12. Again, the empty sum rule yields consistent results as expected.

$$\log 2 = 1 - \gamma + \sum_{k=2}^{\infty}(-1)^k \frac{\zeta(k) - 1}{k} \qquad (2.3.11)$$

$$\log(1+x) = \sum_{k=1}^{\infty}(-1)^{k+1} \frac{H_{k,x}}{k} \qquad (2.3.12)$$

Moreover, we rewrite Eq. 2.3.9 to have:

$$\log(1+x) = -\gamma + H_x + \frac{1}{2}\sum_{k=x+1}^{\infty} 1/k^2 - \frac{1}{3}\sum_{k=x+1}^{\infty} 1/k^3 + \cdots \qquad (2.3.13)$$

Using the series expansion of $\log(1+x)$ for $|x| < 1$, we note that Eq. 2.3.13 can be rearranged as follows:

$$\log(1+x) = -\gamma + H_x + \sum_{k=x+1}^{\infty}\left(\frac{1}{k} - \log\left(1+\frac{1}{k}\right)\right) \qquad (2.3.14)$$

Setting $x = 0$ yields Euler's well-known identity [Weia, Seb02]:

$$\gamma = \sum_{k=1}^{\infty} \left(\frac{1}{k} - \log\left(1 + \frac{1}{k}\right) \right) \qquad (2.3.15)$$

In this example, we have reproduced some of the most basic results that are related to Euler's constant. The fact that all of these results were proven in only a few lines using elementary tools is an example of how effective summability calculus truly is.

2.3.3 The Beta Function

In this example, we look into the beta function. The beta function is given by:

$$B(x, y) = \int_0^1 t^{x-1}(1 - t)^{y-1} dt \qquad (2.3.16)$$

It is symmetric with respect to x and y, and it satisfies the recurrence identity:

$$\log B(x, y) = \log \frac{x - 1}{y + x - 1} + \log B(x - 1, y)$$

Therefore:

$$\log B(n + 1, y) = -\log y + \sum_{k=0}^{n-1} \log \frac{k + 1}{y + k + 1},$$

where we used the initial condition $B(1, y) = \frac{1}{y}$. From this, we have:

$$\log B(n + 1, y) = -\log y + \log n! - \sum_{k=0}^{n-1} \log(y + k + 1) \qquad (2.3.17)$$

We introduce the following function:

$$\gamma(x) = \lim_{s \to \infty} \left\{ \sum_{k=0}^{s-1} \frac{1}{x + k + 1} - \log(x + s + 1) \right\}$$

$$= -\log(1 + x) + \sum_{k=1}^{\infty} \frac{1}{x + k} - \log\left(1 + \frac{1}{x + k}\right)$$

Note that $\gamma(0) = \gamma$ is the Euler constant. Using Theorem 2.2, we have by differentiation:

$$\frac{d}{dn} \log B(n+1, x) = -(\gamma - \gamma(x)) + \sum_{k=0}^{n-1} \frac{1}{k+1} - \frac{1}{x+k+1}$$

Setting $n = 0$ and using the definitions of both the beta function in Eq. 2.3.16 and $\gamma(x)$, we obtain:

$$\gamma(x) = \gamma + x \int_0^1 t^{x-1} \log(1-t)dt$$

$$= -\log(1+x) + \sum_{k=1}^{\infty} \frac{1}{x+k} - \log\left(1 + \frac{1}{x+k}\right)$$

$$= -\log(1+x) + \sum_{k=1}^{\infty} \left(\frac{1}{2(x+k)^2} - \frac{1}{3(x+k)^3} + \frac{1}{4(x+k)^4} - \cdots\right)$$

$$= -\log(1+x) + \sum_{k=2}^{\infty} (-1)^k \frac{\zeta(k, x)}{k}$$

Now, if we expand the integrand in Eq. 2.3.16 into a series expansion and integrate, we obtain:

$$\gamma(x) = \gamma - x\left[\int_0^1 \left(t^x + \frac{t^{x+1}}{2} + \frac{t^{x+2}}{3} + \frac{t^{x+3}}{4} + \cdots\right)dt\right]$$

$$= \gamma - x\left[\frac{1}{1(x+1)} + \frac{1}{2(x+2)} + \frac{1}{3(x+3)} + \frac{1}{4(x+4)} + \cdots\right]$$

So, we have:

$$\gamma(x) = \gamma - x \int_0^1 t^{x-1} \log(1-t)dt$$

$$= \gamma - x\left[\frac{1}{1(x+1)} + \frac{1}{2(x+2)} + \frac{1}{3(x+3)} + \frac{1}{4(x+4)} + \cdots\right]$$

$$= -\log(1+x) + \sum_{k=2}^{\infty} (-1)^k \frac{\zeta(k, x)}{k}$$

This gives us the following characterization of the Euler constant, which holds for all x:

$$\gamma = -\log(1+x) + \sum_{k=2}^{\infty}(-1)^k \frac{\zeta(k,x)}{k} + x\sum_{k=1}^{\infty}\frac{1}{k(x+k)}$$

$$= -\log(1+x) + \sum_{k=2}^{\infty}(-1)^k \frac{\zeta(k,x)}{k} + H_x \qquad (2.3.18)$$

Here, we used a series representation of the harmonic numbers, which we will derive later in Sect. 6.1 (Exercise 6.1). To reiterate, the above expression holds for all $x \in \mathbb{C}$. If we let $x \to \infty$, we recover the original definition of γ in Eq. 2.3.10.

Now, we may combine the above equations with Eq. 5.1.33 that will be proved later in Chap. 5 to obtain:

$$\frac{\log \pi - \gamma}{2} = \sum_{k=0}^{\infty}(-1)^k \Big(H_{k+1} - \log(1+k) - \gamma \Big)$$

$$= \sum_{k=0}^{\infty}(-1)^k \Big(H_k - \log(1+k) - \gamma \Big) + \sum_{k=0}^{\infty}\frac{(-1)^k}{k+1}$$

$$= \log 2 + \sum_{k=0}^{\infty}(-1)^k \Big(H_k - \log(1+k) - \gamma \Big)$$

$$= \log 2 + \sum_{k=0}^{\infty}(-1)^k \sum_{r=2}^{\infty}(-1)^{r+1}\frac{\zeta(r,k)}{r}$$

$$= \log 2 + \sum_{k=0}^{\infty}(-1)^k \sum_{r=1}^{\infty}\Big[\frac{1}{r+k} - \log\Big(1+\frac{1}{r+k}\Big)\Big]$$

Consequently:

$$\sum_{k=0}^{\infty}(-1)^{k+1} \sum_{r=1}^{\infty}\Big[\frac{1}{r+k} - \log\Big(1+\frac{1}{r+k}\Big)\Big] = \frac{\log\frac{\pi}{4} - \gamma}{2}$$

After interchanging the summations and replacing the iteration variable r with $r+1$, one obtains:

$$\sum_{r=1}^{\infty}\sum_{k=0}^{\infty}(-1)^k \Big[\frac{1}{r+k} - \log\Big(1+\frac{1}{r+k}\Big)\Big] = \frac{\log\frac{4}{\pi} + \gamma}{2} \qquad (2.3.19)$$

In this equation, the right-hand side is the average of the two constants $\log \frac{4}{\pi}$ and γ. Note that $\log \frac{4}{\pi}$ has been interpreted as an "alternating" Euler constant in the past and was found to be intimately connected to γ [Son10].

2.3.4   Asymptotic Expansions

Finally, we conclude this section with two simple proofs to the asymptotic behavior of the factorial and the hyperfactorial functions. The asymptotic expressions we derive here are commonly known as the Stirling and the Glaisher approximations respectively [Gla93, Rob55, Weib], and were provided earlier in Eqs. 1.1.12 and 1.1.13.

Our starting point will be Eq. 2.3.9. Replacing x with n and integrating both sides using the integration rule in Proposition 2.3 yields:

$$(1 + n) \log (1 + n) = c_1 + c_2 n + \log n! + \frac{H_n}{2} - \frac{H_{2,n}}{6} + \frac{H_{3,n}}{12} \cdots \qquad (2.3.20)$$

Setting $n = 0$ yields $c_1 = 0$. Setting $n = 1$ and using the identity $2 \log 2 - 1 = \sum_{k=1}^{\infty} (-1)^{k+1}/(k(k+1))$ implies that $c_2 = 1$. Thus, we have:

$$(1 + n) \log (1 + n) - n - \log n! = \frac{H_n}{2} - \frac{H_{2,n}}{6} + \frac{H_{3,n}}{12} \cdots \qquad (2.3.21)$$

Now, knowing that $H_n \sim \log n + \gamma$, we obtain the following asymptotic expression for the log-factorial function, whose error term goes to zero as $n \to \infty$:

$$\log n! \sim (1 + n) \log (1 + n) - n - \frac{\log n}{2} - \frac{\gamma}{2} + \frac{\zeta(2)}{6} - \frac{\zeta(3)}{12} + \cdots \qquad (2.3.22)$$

However, upon using the well-known identity in Eq. 2.3.23, we arrive at the asymptotic formula for the factorial function given in Eq. 2.3.24. Clearly, it is straightforward to arrive at the Stirling approximation from Eq. 2.3.24 but the asymptotic expression in Eq. 2.3.24 is more accurate.

$$-\frac{\gamma}{2} + \frac{\zeta(2)}{6} - \frac{\zeta(3)}{12} + \cdots = \frac{\log 2\pi}{2} - 1 \qquad (2.3.23)$$

$$n! \sim \frac{1}{e} \sqrt{2\pi(n + 1)} \left(\frac{n + 1}{e}\right)^n \qquad (2.3.24)$$

On the other hand, if we start with the log-hyperfactorial function [Weic] $f(n) = \log H(n) = \sum_{k=0}^{n-1} (1 + k) \log (1 + k)$, and differentiate using the differentiation rule for simple finite sums, we obtain:

$$\frac{H'(n)}{H(n)} = n + \log n! + c \qquad (2.3.25)$$

Unfortunately, $\log H(n)$ is not semi-linear so we cannot use Theorem 2.2 to find the exact value of c. These general cases will be addressed in the following section. Nevertheless, since we have the series expansion of the log-factorial function, we can substitute it in the previous equation and integrate with respect to n to arrive at:

$$\log H(n) = \frac{1}{4} + \frac{(1+n)^2(2\log(1+n) - 1)}{2} - \frac{n + \log n!}{2} - \frac{H_n}{6} + \frac{H_{2,n}}{24} - \frac{H_{3,n}}{60} \cdots \tag{2.3.26}$$

Here, we used the facts that $\log H(1) = 0$ and $\log H(0) = 0$ by definition of the hyperfactorial function and the empty sum rule respectively. Thus, an asymptotic expression for the hyperfactorial function is given by:

$$H(n) \sim \frac{(1+n)^{(1+n)^2/2} e^K}{e^{n+n^2/4} n^{1/6} \sqrt{n!}}, \qquad K = -\frac{\gamma}{6} + \frac{\zeta(2)}{24} - \frac{\zeta(3)}{60} + \frac{\zeta(4)}{120} - \cdots \tag{2.3.27}$$

Again, it is straightforward to proceed from the asymptotic expression in Eq. 2.3.27 to arrive at the Glaisher's approximation formula in Eq. 1.1.13 through simple algebraic manipulations but Eq. 2.3.27 is much more accurate.

Exercise 2.4 (Translation Invariance) Prove that the definition of fractional finite sums given by Theorem 2.2 formally satisfies the *translation invariance* property:

$$\sum_{k=0}^{n-1} g(k+r) = \sum_{k=0}^{n-1+r} g(k) - \sum_{k=0}^{r-1} g(k) \doteq \sum_{k=r}^{n-1+r} g(k)$$

This translation invariance property has been proposed as one of six axioms of fractional finite sums in [MS11].

Exercise 2.5 (Telescoping Sums) Prove that any definition of fractional finite sums that satisfies the translation invariance property (see Exercise 2.4) as well as the two defining properties of simple finite sums listed in Chap. 1, must also be consistent with telescoping sums. That is, prove that the three properties imply that if $f(n) = \sum_{k=0}^{n-1} h(k+1) - h(k)$ for some function h, then $f_G(n) = h(n) - h(0)$ for all $n \in \mathbb{C}$.

Exercise 2.6 Employ the techniques used in Sect. 2.3.2 to show that:

$$\sum_{k=1}^{\infty} \frac{\zeta(2k+1)}{4^k(2k+1)} = \log 2 - \gamma \tag{2.3.28}$$

Exercise 2.7 Prove that:

$$\log n! = -\frac{\gamma - 1}{1}(n-1) + \frac{\zeta(2) - 1}{2}(n-1)^2 - \frac{\zeta(3) - 1}{3}(n-1)^3 + \frac{\zeta(4) - 1}{4}(n-1)^4 - \cdots$$

Use this result to derive the following identities:

$$\sum_{k=2}^{\infty}(-1)^k(\zeta(k) - 1) = \frac{1}{2} \qquad (2.3.29)$$

$$\sum_{k=2}^{\infty}(\zeta(k) - 1) = 1 \qquad (2.3.30)$$

$$\sum_{k=1}^{\infty}(\zeta(2k) - 1) = 1 - \frac{\gamma}{2} \qquad (2.3.31)$$

$$\sum_{k=2}^{\infty}\frac{\zeta(k - 1)}{k} = 1 - \gamma \qquad (2.3.32)$$

$$\sum_{k=1}^{\infty}\frac{\zeta(2k + 1) - 1}{2k + 1} = 1 - \gamma - \frac{\log 2}{2} \qquad (2.3.33)$$

It is worth noting that while summability calculus provides a simple approach for deducing such a rich set of identities, some of these identities were proven in the past at different periods of time by different mathematicians using different approaches (see for instance the surveys in [GS04, Seb02]).

Exercise 2.8 Derive the Taylor series expansion of the function $f(n) = \sum_{k=0}^{n-1}\frac{\log(1+k)}{1+k}$ around the origin. Use your result to prove the following two identities:

$$\sum_{r=1}^{\infty}(H_r(\zeta(r + 1) - 1) + \zeta_{r+1}') = 0 \qquad (2.3.34)$$

$$\sum_{r=1}^{\infty}(-1)^r(H_r(\zeta(r + 1) - 1) + \zeta'(r + 1)) = \frac{\log 2}{2}, \qquad (2.3.35)$$

where $\zeta_s' = -\sum_{k=1}^{\infty}\frac{\log k}{k^s}$.

Exercise 2.9 Consider the following semi-linear simple finite sum $f(n) = \sum_{k=0}^{n-1}\frac{1}{k!} = a(n)/n!$, where $a(n) = 0, 1, 4, 15, 64, \ldots$ has the following closed form expression:

$$a(n) = \int_0^{\infty} e^{1+x-e^{\frac{x}{n}}} dx$$

Use the above equation and the differentiation rule to show that:

$$\frac{1}{n!}\int_0^1 x^n(\log x)e^{-x} dx = H_n - \frac{1}{e}\left[\sum_{k=0}^{n-1}\frac{H_n}{k!} - \sum_{k=0}^{n-1}\frac{H_k}{k!} + \sum_{k=0}^{\infty}\frac{H_k}{k!}\right] \qquad (2.3.36)$$

Express this result in terms of the incomplete gamma function:

$$\Gamma^\star(n, x) = \int_0^x t^{n-1} e^{-t} dt$$

Finally, differentiate both sides of Eq. 2.3.36 and use the known expansion of the incomplete gamma function to prove the following identity:

$$\sum_{k=0}^{\infty} \frac{1}{k!} \left(H_{2,k} + H_{1,k}^2 \right) = 2e \sum_{k=0}^{\infty} \frac{(-1)^k}{(k+1)! \cdot (k+1)^2}, \tag{2.3.37}$$

where $H_{s,k} = \sum_{k=0}^{n-1} 1/(1+k)^s$.

2.4   Simple Finite Sums: The General Case

In Sect. 2.2, we derived the rules of summability calculus for semi-linear simple finite sums. Needless to mention, many important finite sums that arise in practice are not semi-linear including, for example, the log-hyperfactorial function discussed earlier. As illustrated in Sect. 2.3.4, it is possible to perform infinitesimal calculus on simple finite sums that are not semi-linear by reducing analysis to the case of semi-linear simple finite sums using a few tricks. For example, the log-hyperfactorial function is not semi-linear but its derivative is, so we could compute the series expansion of its derivative and integrate subsequently using the rules of summability calculus. In principle, therefore, as long as some higher order derivative of a simple finite sum is semi-linear, we could perform infinitesimal calculus, albeit using quite an involved approach. Obviously, this is far from being satisfactory.

In this section we extend the results of Sect. 2.2 to the general case of simple finite sums, which brings to light the celebrated *Euler-Maclaurin summation formula* [Apo99, Pen02, Spi06].

2.4.1   The Euler-Maclaurin Summation Formula

Our starting point will be the successive polynomial approximation method of Theorem 2.1. Here, Theorem 2.1 proves that the following equation holds formally for all functions g:

$$\frac{d}{dn} \sum_{k=0}^{n-1} g(k) = \sum_{k=0}^{\infty} \frac{B_k}{k!} g^{(k)}(n), \tag{2.4.1}$$

where $B_k = (1, -\frac{1}{2}, \frac{1}{6}, 0, \cdots)$ are the Bernoulli numbers. In general, we have:

$$\frac{d^r}{dn^r} \sum_{k=0}^{n-1} g(k) = \sum_{k=0}^{\infty} \frac{B_k}{k!} g^{(k+r-1)}(n) \qquad (2.4.2)$$

Why are these formal expressions useful? To see the answer, we recall that, by the differentiation rule of simple finite sums, we have for some constants c_r:

$$\frac{d^r}{dn^r} \sum_{k=0}^{n-1} g(k) = \sum_{k=0}^{n-1} g^{(r)}(k) + c_r \qquad (2.4.3)$$

To determine the values of c_r, we can now use the formal expression in Eq. 2.4.2 and equate terms:

$$c_r = \sum_{k=0}^{\infty} \frac{B_k}{k!} g^{(k+r-1)}(n) - \sum_{k=0}^{n-1} g^{(r)}(k) \qquad (2.4.4)$$

Of course, Eq. 2.4.4 is only a formal expression; the infinite sum $\sum_{k=0}^{\infty} \frac{B_k}{k!} g^{(k+r-1)}(n)$ may or may not converge. However, if g has a finite polynomial order, then we can treat the right-hand side of Eq. 2.4.4 as an asymptotic expression [Har49], which gives us the values of c_r that we need.

Theorem 2.3 *Let a simple finite sum be given by* $f(n) = \sum_{k=0}^{n-1} g(k)$ *, where* $g(k)$ *is regular at the origin, and let* $f_G(n)$ *be given formally by the following series expansion around* $n = 0$*, where* B_k *are the Bernoulli numbers:*

$$f_G(n) = \sum_{k=0}^{\infty} \frac{c_k}{k!} n^k, \qquad c_k = \sum_{r=0}^{\infty} \frac{B_r}{r!} g^{(r+k-1)}(n) - \sum_{j=0}^{n-1} g^{(k)}(j) \qquad (2.4.5)$$

Here, c_k *are constants that are independent of n. Then,* $f_G(n)$ *satisfies formally the recurrence identity and initial conditions given by Eq. 2.1.1.*

Theorem 2.3 follows from the proof of the successive polynomial approximation method of Theorem 2.1. Alternatively, it can be proved directly in a manner that is similar to the proof of Theorem 2.2, which we leave as an exercise to the reader. The key ingredient is the following identity for the Bernoulli numbers:

$$\sum_{k=0}^{r} \frac{B_r}{k! \, (r-k)!} = 0 \qquad (2.4.6)$$

Remark 2.2 (Polynomial Fitting) We will establish in Chap. 6 that the Euler-Maclaurin summation formula corresponds to a polynomial fitting. In fact, this is already hinted at by the statement of Theorem 2.1. Similar to the earlier argument that summability calculus with semi-linear simple finite sums corresponds to linear fitting, which is perhaps the unique most natural method of generalization, polynomial fitting is arguably the unique most natural generalization of simple finite sums.

Corollary 2.3 (The Euler-Maclaurin Summation Formula) *Let $f(n)$ be a simple finite sum given by $f(n) = \sum_{k=0}^{n-1} g(k)$, where $g(k)$ is regular at the origin, and let $f_G(n)$ be as defined in Theorem 2.3. Then, $f_G(n)$ is given formally by:*

$$f_G(n) = \int_0^n g(t)\,dt + \sum_{r=1}^{\infty} \frac{B_r}{r!}\left(g^{(r-1)}(n) - g^{(r-1)}(0)\right) \qquad (2.4.7)$$

Proof By integrating both sides of Eq. 2.4.1. □

What does Theorem 2.3 imply? Again, if we look into the differentiation rule of simple finite sums in Proposition 2.2, we realize that its definite constant c can be determined once we know, at least, one value of $f_G'(n)$ for some $n \in \mathbb{N}$. This is because we always have $c = f_G'(n) - \sum_{k=0}^{n-1} g'(k)$ independently of n. In the case of semi-linear simple finite sums, in which $g'(k)$ vanishes as $k \to \infty$, the rate of change of the simple finite sum becomes almost linear when $n \to \infty$, so we know that the derivative has to be arbitrarily close to $g(n)$ as $n \to \infty$. This was, in fact, the value we were looking for so we used it to determine the definite value of c. Theorem 2.2 establishes that such an approach is correct.

In Theorem 2.3, on the other hand, we look into the general case of simple finite sums in which $g(k)$ is not necessarily a nearly-convergent function. In this case, we still need to find, at least, one value of $f_G'(n)$ for some $n \in \mathbb{N}$ so that the constant c in Eq. 2.4.3 can be determined. Equation 2.4.2 provides us with a formal answer. However, because the infinite sum typically diverges, we still need a way of interpreting such a divergent series.

In Chaps. 4 and 6, we will present a new summability method χ, which works reasonably well in estimating the Euler-Maclaurin summation formula even if it diverges. In the meantime, we note that the Euler-Maclaurin summation formula is an asymptotic expression; we can clip the infinite sum *if $g(k)$ has a finite polynomial order* (see Definition 1.1). If g has a finite polynomial order $r < \infty$, we have the asymptotic expression:

$$f_G'(n) \sim \sum_{k=0}^{r} \frac{B_k}{k!} g^{(k)}(n)$$

Note that because $g^{(r+1)}(n) \to 0$ as $n \to \infty$, we do *not* have to evaluate the entire Euler-Maclaurin summation formula. The last *finite* asymptotic expression is all the information we need in order to compute the value of c, in a manner that is similar to what was done earlier for semi-linear simple finite sums. Thus, we have:

$$f_G^{(r)}(0) = \lim_{n \to \infty} \left\{ \sum_{k=0}^{r} \frac{B_k}{k!} g^{(k+r-1)}(n) - \sum_{k=0}^{n-1} g^{(r)}(k) \right\} \tag{2.4.8}$$

Of course, one notable difference between Theorems 2.3 and 2.2, however, is that we could find the constants c_r in Theorem 2.3 by choosing *any appropriate value of n*, at least in principle, whereas we had to take the limit as $n \to \infty$ in Theorem 2.2 and in Eq. 2.4.8. Nevertheless, Eq. 2.4.8 is usually more useful in practice. We will illustrate both approaches in Sect. 2.5.

Remark 2.3 It is straightforward to observe that Theorem 2.3 generalizes Theorem 2.2 that was proved earlier for semi-linear finite sums.

Exercise 2.10 Prove that the Euler-Maclaurin summation formula is consistent with the differentiation rule of Proposition 2.2. In other words, prove that:

$$\frac{d}{dn} \sum_{k=0}^{n-1} g(k) = \sum_{k=0}^{n-1} g'(k) + c,$$

for some constant c that is independent of n, if $\sum_{k=0}^{n-1} g(k)$ is defined for all $n \in \mathbb{C}$ by Corollary 2.3.

Exercise 2.11 Prove that the Euler-Maclaurin summation formula satisfies the following two properties:

1. *Translation Invariance*: $\sum_{k=0}^{n-1} g(k+r) = \sum_{k=0}^{n-1+r} g(k) - \sum_{k=0}^{r-1} g(k)$
2. *Linearity*: $\sum_{k=0}^{n-1} (\alpha g(k) + h(k)) = \alpha \sum_{k=0}^{n-1} g(k) + \sum_{k=0}^{n-1} h(k)$.

Exercise 2.12 Using the fact that $\frac{x}{e^x - 1} = \sum_{k=0}^{\infty} \frac{B_k}{k!} x^k$, prove that the Bernoulli numbers satisfy the identity:

$$\sum_{k=0}^{n-1} \binom{n}{k} B_k = 0$$

This is the identity presented in Eq. 2.4.6. (Hint: multiply by $e^x - 1$ and equate coefficients)

2.5 Examples to the General Case of Simple Finite Sums

In this section, we present a few examples for performing infinitesimal calculus on simple finite sums using Theorem 2.3 and its corollaries.

2.5.1 Beyond the Bernoulli-Faulhaber Formula

In this example, we return to the power sum function given by Eq. 1.1.5. If m is a positive integer, we can use Theorem 2.3 to deduce immediately the Bernoulli-Faulhaber formula for power sums, which is the original proof that Euler provided. However, the Bernoulli-Faulhaber formula does not provide us with a series expansion if m is an arbitrary positive real number. Using Theorem 2.3, on the other hand, we can quite easily derive a series expansion of the power sum function $f(n; m) = \sum_{k=0}^{n-1}(1 + k)^m$ around the origin, which extends the Bernoulli-Faulhaber formula.

First, we note that $f(0) = 0$ by the empty sum rule and that $g^{(r)}(n) = \frac{m!}{(m-r)!}(1 + n)^{m-r}$. Since $\lim_{n\to\infty} g^{(r)}(n) = 0$ for $r > m$, we can choose $n \to \infty$ in Theorem 2.3 to find the values of the derivatives $c_r = f_G^{(r)}(0)$ in Eq. 2.4.3. We illustrate this procedure when $m = 1/2$.

First, we note that:

$$c_1 = \lim_{n\to\infty} \left\{ \sqrt{1 + n} - \sum_{k=0}^{n-1} \frac{1}{2\sqrt{1 + k}} \right\} \approx 0.7302 \qquad (2.5.1)$$

Later in Chap. 5, we provide the machinery to deduce that the constant c_1 is, in fact, given by $-\frac{\zeta(1/2)}{2}$, where $\zeta(s)$ is the Riemann zeta function. Using the differentiation rule:

$$f_G'(n) = -\frac{\zeta(1/2)}{2} + \sum_{k=0}^{n-1} \frac{1}{2\sqrt{1 + k}} \qquad (2.5.2)$$

Similarly:

$$c_2 = \sum_{k=1}^{\infty} \frac{1}{4k^{3/2}} = \frac{\zeta(3/2)}{4} \qquad (2.5.3)$$

Continuing in this manner, we note that:

$$c_r = (-1)^r \frac{(2m - 3)!}{4^{r-1}(r - 2)!} \zeta(r - 1/2), \qquad r \geq 0 \qquad (2.5.4)$$

Thus, the series expansion is given by:

$$f_G(n) = \sum_{k=0}^{n-1} \sqrt{1 + k} = -\frac{\zeta(1/2)}{2} n + \sum_{r=2}^{\infty} (-1)^r \frac{(2r - 3)!}{4^{r-1} r! (r - 2)!} \zeta(r - 1/2) n^r$$

$$(2.5.5)$$

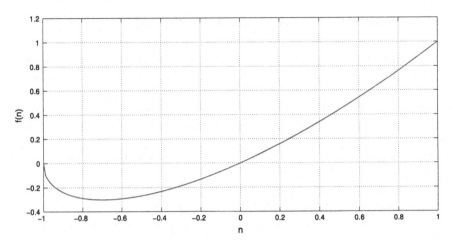

Fig. 2.2 The function $\sum_{k=1}^{n} \sqrt{k}$ plotted in the interval $-1 \le n \le 1$

Unfortunately, the radius of converges of this series is $|n| < 1$, which can be deduced using the Stirling approximation. However, we can still perform a sanity check on Eq. 2.5.5 by plotting it for $-1 \le n \le 1$, which is shown in Fig. 2.2. As shown in the figure, $f(0) = 0$ and $f(1) = 1$ as expected. Later in Chap. 4, we will introduce a summability method χ, which can be used to numerically verify that the Taylor series expansion derived in Eq. 2.5.5 satisfies $f_G(2) = \sqrt{1} + \sqrt{2}$ and $f_G(3) = \sqrt{1} + \sqrt{2} + \sqrt{3}$, thus confirming our previous analysis.

Earlier, we stated that Theorem 2.3 provides a more convenient approach for performing infinitesimal calculus on simple finite sums than the classical Euler-Maclaurin summation formula. To see this in our example, it is worthwhile to note that the Euler-Maclaurin summation formula does not readily extend the domain of $\sum_{k=0}^{n-1} \sqrt{1+k}$ to negative values of n nor can it be used in deriving the Taylor series expansion because of its divergence. By contrast, Theorem 2.3 provides us with a simple method for achieving these feats. Finally, similar techniques can be used to obtain the Taylor series expansions for other values of m.

2.5.2   The Superfactorial Function

In this example, we look into the simple finite sum $f(n) = \sum_{k=0}^{n-1} \frac{d}{dk} \log(k!)$. The finite sum $f(n)$ is semi-linear so we can use Theorem 2.2 directly. Using the differentiation rule, we have:

$$f'_G(n) = c + \sum_{k=0}^{n-1} \frac{d^2}{dk^2} \log k! \qquad (2.5.6)$$

Using Theorem 2.2, the value of c is given by:

$$c = \lim_{n \to \infty} \left\{ -\gamma + H_n - \sum_{k=0}^{n-1} (\zeta(2) - H_{2,k}) \right\} = -(1 + \gamma) \qquad (2.5.7)$$

Continuing in this manner yields the following series expansion for $f_G(n)$:

$$f_G(n) = \sum_{k=0}^{n-1} (-\gamma + H_k) = -(1 + \gamma)n + \sum_{k=2}^{\infty} (-1)^k \zeta(k) \, n^k \qquad (2.5.8)$$

By comparing the series expansion in Eq. 2.5.8 with the series expansion of $\log n!$ that was derived in Eq. 2.3.3, we deduce that:

$$\sum_{k=0}^{n-1} (-\gamma + H_k) = n\left(-\gamma + H_n - 1 \right) \qquad (2.5.9)$$

From this, we conclude that:

$$\sum_{k=0}^{n-1} H_k = n(H_n - 1) \qquad (2.5.10)$$

We are now ready to find the series expansion of the log-superfactorial function $\log S(n) = \sum_{k=0}^{n-1} \log k!$. By integrating both sides of Eq. 2.5.8 using the integration rule of simple finite sums, we have:

$$\sum_{k=0}^{n-1} \log k! = c_0 + c_1 n - \frac{1 + \gamma}{2} n^2 + \sum_{k=2}^{\infty} (-1)^k \frac{\zeta(k)}{k+1} \, n^{k+1} \qquad (2.5.11)$$

By setting $n = 0$ and the empty sum rule, we obtain $c_0 = 0$. Using $n = 1$, on the other hand, yields:

$$c_1 = \frac{1 + \gamma}{2} - \sum_{k=2}^{\infty} (-1)^k \frac{\zeta(k)}{k+1} = 0.4189 \cdots \qquad (2.5.12)$$

To find an analytic expression for c_1, we employ Theorem 2.3 this time, which states that:

$$c_1 = \lim_{n \to \infty} \left\{ \log n! + \frac{\gamma - H_n}{2} + \sum_{k=0}^{n-1} (\gamma - H_k) \right\}$$

$$= \lim_{n \to \infty} \left\{ \frac{\log 2\pi n}{2} + n(\log n - 1) + \frac{\gamma - H_n}{2} + n\gamma - \sum_{k=0}^{n-1} H_k \right\}$$

Here, we used the Stirling approximation. Upon using Eq. 2.5.9, we have:

$$c_1 = \frac{\log 2\pi}{2} + \lim_{n \to \infty} \left\{ n\left(\log n + \gamma - H_n\right) \right\} \qquad (2.5.13)$$

We know, however, using the Euler-Maclaurin summation formula that the following asymptotic relation holds:

$$\log n + \gamma - H_n \sim -\frac{1}{2n} \qquad (2.5.14)$$

Plugging Eq. 2.5.14 into Eq. 2.5.13 yields the desired value:

$$c_1 = \frac{\log 2\pi - 1}{2} \qquad (2.5.15)$$

Equating Eq. 2.5.15 with Eq. 2.5.12, we arrive at the following identity:

$$\frac{\log 2\pi - 1}{2} = \frac{1 + \gamma}{2} - \sum_{k=2}^{\infty} (-1)^k \frac{\zeta(k)}{k+1} \qquad (2.5.16)$$

This identity was proved independently by various authors [CS97]. Finally, by plugging Eq. 2.5.15 into Eq. 2.5.11, we arrive at the desired series expansion of the log-superfactorial function:

$$\sum_{k=0}^{n-1} \log k! = \frac{\log 2\pi - 1}{2} n - \frac{1 + \gamma}{2} n^2 + \sum_{k=2}^{\infty} (-1)^k \frac{\zeta(k)}{k+1} n^{k+1} \qquad (2.5.17)$$

Equation 2.5.17 has appeared previously in various sources, such as in [CS97] and the references therein.

Exercise 2.13 Explain why the empty sum rule implies that $f(-1) = 0$ when $f(n) = \sum_{k=0}^{n-1} \sqrt{1+k}$ (Hint: use the recurrence identity and the fact that $g(-1) = 0$). Use this fact and the results of Sect. 2.5.1 to prove that:

$$-\frac{1}{2}\zeta(1/2) = \sum_{m=2}^{\infty} \frac{(2m-3)!}{4^{m-1} m! (m-2)!} \zeta(m - 1/2) \qquad (2.5.18)$$

Exercise 2.14 Use Theorem 2.3 and the Stirling approximation to prove that the Taylor series expansion of the log-hyperfactorial function $f(n) = \sum_{k=0}^{n-1} (1 + k) \log(1 + k)$ is given by:

$$\sum_{k=0}^{n-1} (1 + k) \log(1 + k) = \left(\frac{1 - \log 2\pi}{2}\right) n + \frac{1 - \gamma}{2} n^2 + \sum_{k=2}^{\infty} (-1)^k \frac{\zeta(k)}{k(k+1)} n^{k+1}$$

$$\qquad (2.5.19)$$

Exercise 2.15 Use *summation by parts* to prove that the harmonic numbers satisfy the identity $\sum_{k=0}^{n-1} H_k = n(H_n - 1)$, which was proved in Eq. 2.5.9 using the differentiation rule of simple finite sums. Summation by parts states that for any functions g and h, we have:

$$\sum_{k=0}^{n-1} g(k)\big[h(k+1) - h(k)\big] = g(n)h(n) - g(0)h(0) - \sum_{k=0}^{n-1} h(k+1)\big[g(k+1) - g(k)\big]$$

$$(2.5.20)$$

Exercise 2.16 Use the differentiation rule of simple finite sums and the empty sum rule to prove that for $f(n) = \sum_{k=0}^{n-1} e^{i\pi k}$, we have $f_G^{(r)}(0) = (i\pi)^{r-1} f_G'(0)$. Use this result to show that:

$$\sum_{k=0}^{n-1} e^{i\pi k} = \frac{1 - e^{i\pi n}}{2} \qquad\qquad (2.5.21)$$

Compare this with the solution that would have been obtained using the well-known closed-form expression of geometric progressions of the form $\sum_{k=0}^{n-1} x^k$. Use the Euler-Maclaurin summation formula to derive, again, the same results.

Exercise 2.17 Use the results of Sect. 2.5.2 to prove the following identity:

$$\int_0^n \log \Gamma(t+1)dt = \frac{1}{2}(\log(2\pi) - 1)\,n - \frac{n^2}{2} + n\,\log n! - \sum_{k=0}^{n-1} \log k! \qquad (2.5.22)$$

This identity is sometimes referred to as Alexeiewsky's theorem [CS97]. (Hint: begin with Eq. 2.5.9 and use the integration rule of simple finite sums)

Remark 2.4 The approach used in Exercise 2.16 for oscillating simple finite sums is not very insightful. For instance, it is difficult to use the Euler-Maclaurin summation formula to derive the asymptotic expansions to oscillating finite sums, due to the presence of complex values. We will resolve this difficulty later when we derive the analog of the Euler-Maclaurin summation formula for oscillating finite sums in Chap. 5, which will allow us to perform many deeds with ease.

2.6 Summary

In this chapter, the foundational rules of summability calculus are derived, which address the case of simple finite sums.

There are many questions that remain unanswered. For example, the results that have been derived so far do not address the general case of composite finite sums, which are of the form $\sum_{k=0}^{n-1} g(k, n)$. Moreover, the only approach that has been presented so far for *computing* fractional finite sums relies on deriving the Taylor series expansions. Is there a more direct approach for computing fractional finite sums? Even more, the approach presented so far does not provide a convenient method for handling oscillating finite sums. These questions will be answered in the remaining chapters.

References

[Apo99] T.M. Apostol, An elementary view of Euler's summation formula. Am. Math. Mon. **106**(5), 409–418 (1999)

[Blo93] D.M. Bloom, An old algorithm for the sums of integer powers. Math. Mag. **66**, 304–305 (1993)

[CS97] J. Choi, H.M. Srivastava, Sums associated with the zeta function. J. Math. Anal. Appl. **206**(1), 103–120 (1997)

[Dav59] P.J. Davis, Leonhard Euler's integral: a historical profile of the gamma function. Am. Math. Mon. **66**(10), 849–869 (1959)

[Gla93] J. Glaisher, On certain numerical products in which the exponents depend upon the numbers. Messenger Math. **23**, 145–175 (1893)

[GS04] X. Gourdon, P. Sebah, The Euler constant (2004). http://numbers.computation.free.fr/ Constants/Gamma/gamma.html. Retrieved on June 2012

[Har49] G.H. Hardy, *Divergent Series* (Oxford University Press, New York, 1949)

[Hav03] J. Havil, *Gamma: Exploring Euler's Constant* (Princeton University Press, Princeton, 2003)

[Kra99] S.G. Krantz, The Bohr-Mollerup theorem, in *Handbook of Complex Variables* (Springer, New York, 1999), p. 157

[MS11] M. Müller, D. Schleicher, How to add a noninteger number of terms: from axioms to new identities. Am. Math. Mon. **118**(2), 136–152 (2011)

[Pen02] D.J. Pengelley, Dances between continuous and discrete: Euler's summation formula, in *Proceedings of Euler 2K+2 Conference* (2002)

[Rob55] H. Robbins, A remark on Stirling's formula. Am. Math. Mon. **62**, 26–29 (1955)

[Ros80] K.A. Ross, Weierstrass's approximation theorem, in *Elementary Analysis: The Theory of Calculus* (Springer, New York, 1980), p. 200

[Seb02] P. Sebha, Collection of formulae for Euler's constant (2002). http://scipp.ucsc.edu/~haber/ archives/physics116A06/euler2.ps. Retrieved on March 2011

[Shi07] S.A. Shirali, On sums of powers of integers. Resonance **12**, 27–43 (2007)

[Son10] J. Sondow, *New Vacca-Type Rational Series for Euler's Constant γ and Its 'Alternating' Analog* (Springer, New York, 2010), pp. 331–340

[Spi06] M.Z. Spivey, The Euler-Maclaurin formula and sums of powers. Math. Mag. **79**(1), 61–65 (2006)

[TP85] M. Tennenbaum, H. Pollard, *Ordinary Differential Equations: An Elementary Textbook for Students of Mathematics, Engineering, and the Sciences* (Dover Publications, Inc., New York, 1985), p. 91

[Weia] E. Weisstein, Euler-Mascheroni constant. http://mathworld.wolfram.com/Euler-MascheroniConstant.html

[Weib] E. Weisstein, Glaisher-Kinkelin constant. http://mathworld.wolfram.com/Glaisher-KinkelinConstant.html

[Weic] E. Weisstein, Hyperfactorial. http://mathworld.wolfram.com/Hyperfactorial.html

[Weid] E. Weisstein, Stieltjes constants. http://mathworld.wolfram.com/StieltjesConstants.html

Chapter 3
Composite Finite Sums

A complex system that works is invariably found to have evolved from a simple system that worked.

John Gall in *Systemantics*

Abstract In this chapter, we extend the previous results of Chap. 2 to the more general case of composite finite sums. We describe what composite finite sums are and how their analysis can be reduced to the analysis of simple finite sums using the chain rule. We apply these techniques, next, on numerical integration and on some identities of Ramanujan.

In this chapter, we extend the previous results of Chap. 2 to the case of composite finite sums.

3.1 Infinitesimal Calculus on Composite Finite Sums

3.1.1 The Chain Rule

As discussed earlier, composite finite sums are a generalization to simple finite sums in which the iterated function g is itself a function of the bound n. Thus, instead of dealing with finite sums of the form $f(n) = \sum_{k=0}^{n-1} g(k)$, we now have to deal with finite sums of the form $f(n) = \sum_{k=0}^{n-1} g(k, n)$.

Such composite sums are markedly different. For instance, the fundamental recurrence identity in simple finite sums is no longer true, and the two defining properties of simple finite sums that we started with are no longer valid, either. However, it is perhaps one remarkable result of summability calculus that composite finite sums are, in fact, as straightforward to analyze as simple finite sums. This includes, for instance, performing infinitesimal calculus, deducing asymptotic behavior, as well as computing composite finite sums at fractional values of n. The key tool that we will employ is the classical *chain rule* of calculus.

© Springer International Publishing AG 2018
I. M. Alabdulmohsin, *Summability Calculus*,
https://doi.org/10.1007/978-3-319-74648-7_3

Lemma 3.1 (The Chain Rule) *Given a composite finite sum* $f(n) = \sum_{k=0}^{n-1} g(k, n)$, *let* $h(x, y) = \sum_{k=0}^{y-1} g(k, x)$. *Then:*

$$\frac{d}{dn} \sum_{k=0}^{n-1} g(k, n) = \frac{\partial h}{\partial x}\Big|_{x=n} + \frac{\partial h}{\partial y}\Big|_{y=n}$$

$$= \sum_{k=0}^{n-1} \frac{\partial}{\partial n} g(k, n) + \frac{d}{dn} \sum_{k=0}^{n-1} g(k, x)\Big|_{x=n}$$

In other words, the derivative of a composite finite sum decomposes into the sum of two parts: (1) a finite sum of derivatives, and (2) a derivative of a *simple* finite sum. The first part can be computed using ordinary calculus. The second part can be determined using the earlier results of Chap. 2.

3.1.2 Example

Let us illustrate how the above procedure works on the composite finite sum:

$$f(n) = \sum_{k=0}^{n-1} \frac{1}{1+k+n}$$

In this example, we will also show how the aforementioned method is indeed consistent with the earlier results of summability calculus for simple finite sums.

Now, given the definition of $f(n)$, suppose we would like to find its derivative at $n = 1$. To do this, the first component in the chain rule of Lemma 3.1 is given by:

$$\sum_{k=0}^{0} \frac{\partial}{\partial n} \frac{1}{1+k+n} = \frac{d}{dn} \frac{1}{1+n} = -\frac{1}{(1+n)^2}\Big|_{n=1} = -\frac{1}{4}$$

For the second component of the chain rule, we would like to differentiate the simple finite sum $\sum_{k=0}^{n-1} \frac{1}{2+k}$ (note here that we have substituted the value of $n = 1$ inside the iterated function $g(k, n)$). Clearly, the derivative in the latter case is given by:

$$\frac{d}{dn} \sum_{k=0}^{n-1} \frac{1}{2+k} = \sum_{k=n}^{\infty} \frac{1}{(2+k)^2}\Big|_{n=1} = \zeta(2) - \frac{5}{4}$$

Here, we used the rules of summability calculus for simple finite sums. In particular, we used Theorem 2.2 for semi-linear simple finite sums. So, by the chain rule, we add the last two results to deduce that:

$$\frac{d}{dn}\sum_{k=0}^{n-1}\frac{1}{1+k+n}\bigg|_{n=1} = \zeta(2) - \frac{3}{2}$$

How do we know that this is indeed correct? Or to put it more accurately, why do we know that this corresponds to the derivative of the unique most natural generalization to the composite finite sum?

To answer this question, we note in this example that we have $f(n) = H_{2n} - H_n$, where $H_n = \sum_{k=0}^{n-1} 1/(1 + k)$ is the n-th harmonic number. Therefore, if we were indeed dealing with the unique most natural generalization to such a composite finite sum, this generalization should be consistent with the unique most natural generalization of the harmonic numbers. To test this hypothesis, we differentiate the expression $f(n) = H_{2n} - H_n$ at $n = 1$, which yields by the results of Sect. 2.2:

$$\frac{d}{dn}H_{2n} - H_n\bigg|_{n=1} = 2\Big(\zeta(2) - \sum_{k=0}^{2n-1}\frac{1}{(1+k)^2}\Big) - \Big(\zeta(2) - \sum_{k=0}^{n-1}\frac{1}{(1+k)^2}\Big)\bigg|_{n=1} = \zeta(2) - \frac{3}{2}$$

Thus, the two results agree as expected.

3.1.3 General Results

We will now prove the general statements related to composite finite sums.

Lemma 3.2 (Differentiation Rule of Composite Sums) *Let $f(n) = \sum_{k=0}^{n-1} g(k, n)$ be a composite sum, and let its unique natural generalization to the complex plane $\mathbb{C}$ be denoted $f_G(n)$. Then:*

$$f_G'(n) = \sum_{k=0}^{n-1}\frac{\partial}{\partial n}g(k, n) + \sum_{r=0}^{\infty}\frac{B_r}{r!}\frac{\partial^r}{\partial m^r}g(m, n)\bigg|_{m=n} \tag{3.1.1}$$

Proof By direct application of the chain rule in Lemma 3.1 and Eq. 2.4.2. □

Remark 3.1 While it is true that Lemma 3.2 follows directly from the chain rule and Eq. 2.4.2, how do we know that it corresponds to the derivative of the unique most natural generalization of $f(n)$? The answer to this question rests on the earlier claims for simple finite sums. In particular, the chain rule implies that $f_G'(n)$ is a sum of two components. The first component, which is a finite sum of derivatives, is well-defined and is not impacted by our definition of fractional finite sums. The second component, on the other hand, depends on how the derivative of *simple* finite sums is defined. Because we used the differentiation rule for simple finite sums given in Theorem 2.3 that was shown earlier to correspond to the successive polynomial approximation method, it follows, therefore, that Lemma 3.2 yields the derivative of the unique most natural generalization for composite finite sums as well. In other words, we *reduced* the analysis of composite finite sums into the analysis of simple finite sums using the chain rule.

Unfortunately, Lemma 3.2 does not permit us to derive the asymptotic behavior of composite finite sums. This can be achieved by extending the Euler-Maclaurin summation formula [Apo99, Pen02, Spi06] to composite sums as shown next.

Theorem 3.1 *If the derivative of a function $f_G(n)$ is given by Lemma 3.1 and $g(\cdot; n)$ is regular at the origin with respect to its first argument, then $f_G(n)$ is formally given by:*

$$f_G(n) = \int_0^n g(t, n)\, dt + \sum_{r=1}^{\infty} \frac{B_r}{r!} \left(\frac{\partial^{r-1}}{\partial m^{r-1}} g(m, n) \Big|_{m=n} - \frac{\partial^{r-1}}{\partial m^{r-1}} g(m, n) \Big|_{m=0} \right)$$

In addition, the empty-sum rule $f_G(0) = 0$ continues to hold if $g(\cdot, 0)$ is regular at the origin with respect to its first argument.

Proof First, let us introduce the following function:

$$h(n; r) = \sum_{k=0}^{n-1} \frac{\partial^r}{\partial n^r} g(k, n) \tag{3.1.2}$$

We start from Lemma 3.2 and formally integrate both sides with respect to n as follows:

$$f_G(n) = h(0; 0) + \sum_{r=0}^{\infty} \frac{B_r}{r!} \int_0^n \frac{\partial^r}{\partial m^r} g(m, t) \Big|_{m=t} dt + \int_0^n \sum_{k=0}^{t-1} \frac{\partial}{\partial t} g(k, t)\, dt, \tag{3.1.3}$$

However, $\sum_{k=0}^{t-1} \frac{\partial}{\partial t} g(k, t)$ can also be expressed using Eq. 3.1.3 as follows:

$$\sum_{k=0}^{t-1} \frac{\partial}{\partial t} g(k, t) = h(0; 1) + \sum_{r=0}^{\infty} \frac{B_r}{r!} \int_0^t \frac{\partial^{r+1}}{\partial m^r \partial t_2} g(m, t_2) \Big|_{m=t_2} dt_2$$

$$+ \int_0^t \sum_{k=0}^{t_2-1} \frac{\partial^2}{\partial t_2^2} g(k, t_2)\, dt_2, \tag{3.1.4}$$

We now plug Eq. 3.1.4 into Eq. 3.1.3. Repeating the same process indefinitely and by using Cauchy's formula for repeated integration [Wei], we have:

$$f_G(n) = \sum_{r=0}^{\infty} \frac{h(0; r)}{r!} n^r + \sum_{r=0}^{\infty} \frac{B_r}{r!} \frac{\partial^r}{\partial m^r} \int_0^n \sum_{b=0}^{\infty} \frac{(n-t)^b}{b!} \frac{\partial^b}{\partial t^b} g(m, t) \Big|_{m=t} dt \tag{3.1.5}$$

Now, we use Taylor's theorem, which formally states that:

$$\sum_{b=0}^{\infty} \frac{(n-t)^b}{b!} \frac{\partial^b}{\partial t^b} g(m,t) = g(m,n) \tag{3.1.6}$$

Here, the last identity holds formally because the summation is a Taylor series expansion on the second argument of g around t. Plugging this into Eq. 3.1.5 gives us:

$$f_G(n) = \sum_{r=0}^{\infty} \frac{h(0;r)}{r!} n^r + \sum_{r=0}^{\infty} \frac{B_r}{r!} \int_0^n \frac{\partial^r}{\partial m^r} g(m,n)\Big|_{m=t} dt \tag{3.1.7}$$

Now, if we expand the first term using the definition of h and formally interchange the limit with the summation, we obtain:

$$\sum_{r=0}^{\infty} \frac{h(0;r)}{r!} n^r = \lim_{s\to 0}\left\{ \sum_{k=0}^{s-1} g(k,n+s) \right\} = \sum_{k=0}^{-1} g(k,n) = 0,$$

where we used linearity (see Exercise 2.11) and the empty sum rule.

Finally, by formally expanding the last term in Eq. 3.1.7, we obtain the statement of the theorem. ∎

Remark 3.2 Theorem 3.1 states that if we have a composite finite sum $f(n) = \sum_{k=0}^{n-1} g(k,n)$, then a sufficient condition for the empty sum rule to hold is for $g(\cdot, 0)$ to be analytic at the origin with respect to its first argument. We can see that this condition is, in fact, necessary. If we let $f(n) = \sum_{k=0}^{n-1} \binom{n-1}{k}$, then $f(n) = 2^{n-1}$ holds for all $n \in \mathbb{N}$. One may conjecture that its generalization $f_G(n)$ that is given by Theorem 3.1 is itself 2^{n-1}. However, if this conjecture is true, we have $f_G(0) \neq 0$. The reason the empty sum rue is violated is because $g(\cdot; 0)$ is undefined because the factorial function is singular at $n = -1$. If we correct this by using the different function $f(n) = \sum_{k=0}^{n-1} \binom{n}{k}$, then $f(n) = 2^n - 1$. Here, $g(\cdot; 0)$ is analytic at the origin and the empty sum rule $f(0) = 0$ indeed holds in this case.

Exercise 3.1 Show that if $g(k,n)$ is independent of n, then Theorem 3.1 reduces to the classical Euler-Maclaurin summation formula. In other words, prove that the results derived in this chapter for composite finite sums generalize the earlier results derived for simple finite sums in the previous chapter.

Exercise 3.2 Let $f(n) = \sum_{k=0}^{n-1} g(k+n)$. Let $f_G(n)$ be the unique generalization of $f(n)$ to the complex plane $\mathbb{C}$ that is given by Theorem 3.1. Let $h_G(n)$ be the unique generalization of the *simple* finite sum $f(n) = \sum_{k=0}^{n-1} g(k)$ that is given by Corollary 2.3. Prove that $f_G(n) = h_G(2n) - h_G(n)$.

Exercise 3.3 Prove that for all $n \in \mathbb{C}$, we have $\sum_{k=0}^{n-1} g(n-k) = \sum_{k=0}^{n-1} g(k+1)$. (Hint: you may find the following fact useful: $B_{2n+1} = 0$ for all $n > 1$.)

3.2   Examples to Composite Finite Sums

3.2.1   Numerical Integration

The classical approach for approximating definite integrals uses Riemann sums. In this example, we show how Theorem 3.1 yields higher-order approximations, which, in turn, can be used to derive one of the well-known recursive definitions of Bernoulli numbers.

First, we start with the following equation that follows by a direct application of Theorem 3.1:

$$\frac{m}{n}\sum_{k=0}^{n-1} f(x_0 + \frac{m}{n}k) = \int_{x_0}^{x_0+m} f(t)\,dt + \sum_{r=1}^{\infty} \frac{B_r}{r!}\left(\frac{m}{n}\right)^r \left(f^{(r-1)}(x_0+m) - f^{(r-1)}(x_0)\right)$$

Now, we let $n = \frac{x-x_0}{\Delta x}$ and let $m = x - x_0$. After plugging these two expressions into the last equation and rearranging the terms, we arrive at:

$$\int_{x_0}^{x} f(t)\,dt = \sum_{k=0}^{\frac{x-x_0}{\Delta x}-1} f(x_0 + k\Delta x)\Delta x - \sum_{r=1}^{\infty} \frac{B_r}{r!}\left(f^{(r-1)}(x) - f^{(r-1)}(x_0)\right)(\Delta x)^r$$

Clearly, the first term on the right-hand side is the classical approximation of definite integrals, but adding additional terms yields higher-order approximations. In particular, if we choose $\Delta x = x - x_0$, we obtain:

$$\int_{x_0}^{x} f(t)\,dt = f(x_0)\cdot(x - x_0) - \sum_{r=1}^{\infty} \frac{B_r}{r!}\left(f^{(r-1)}(x) - f^{(r-1)}(x_0)\right)(x - x_0)^r \qquad (3.2.1)$$

Using the last equation, we can equate the coefficients of the formal power series in both sides, which yields the following identity of the Bernoulli numbers:

$$\frac{1}{s+1} = -\sum_{r=1}^{s}\binom{s}{r}\frac{B_r}{s-r+1} \qquad (3.2.2)$$

Equation 3.2.2, in turn, can be rearranged to yield the following well-known recursive definition:

$$B_s = -\sum_{r=0}^{s-1}\binom{s}{r}\frac{B_r}{s-r+1} \qquad (3.2.3)$$

3.2.2 An Identity of Ramanujan

In one of his earliest works, Ramanujan was interested in simple finite sums of the form:

$$\phi(x,n) = \sum_{k=0}^{n-1} \frac{1}{(x(1+k))^3 - x(1+k)}$$

He showed that many composite sums can be converted into simple finite sums using properties of the function $\phi(x,n)$ [Ber85]. One result that he derived is the following identity:

$$\sum_{k=0}^{n-1} \frac{1}{1+k+n} = \frac{n}{2n+1} + \sum_{k=0}^{n-1} \frac{1}{8(1+k)^3 - 2(1+k)} \tag{3.2.4}$$

With the aid of the rules of summability calculus for composite finite sums, we can use Eq. 3.2.4 as a starting point to derive many interesting identities. For instance, if we differentiate both sides with respect to n, we obtain:

$$\sum_{k=1}^{n} \frac{1}{k^2} - 2\sum_{k=1}^{2n} \frac{1}{k^2} + \zeta(2) - \frac{n+1}{(2n+1)^2} = \sum_{n+1}^{\infty} \frac{24k^2 - 2}{(8k^3 - 2k)^2} \tag{3.2.5}$$

Selecting $n = 0$, we have the following series for $\zeta(2)$:

$$\zeta(2) = 1 + \sum_{1}^{\infty} \frac{24k^2 - 2}{(8k^3 - 2k)^2} \tag{3.2.6}$$

On the other hand, if we integrate both sides of Eq. 3.2.4 and rearrange the terms, we arrive at:

$$\sum_{k=1}^{n} \log\left(1 - \frac{1}{4k^2}\right) = 2\log(2n)! - 4\log n! + \log(2n+1) - n\log 8 \tag{3.2.7}$$

Taking the limit as $n \to \infty$ yields by Stirling's approximation [Rob55]:

$$\sum_{k=1}^{\infty} \log\left(1 - \frac{1}{4k^2}\right) = -\log\frac{\pi}{2} \tag{3.2.8}$$

Thus, we arrive at a restatement of Wallis formula [SW]:

$$\prod_{k=1}^{\infty}\left(1 - \frac{1}{4k^2}\right) = \frac{2}{\pi} \tag{3.2.9}$$

Exercise 3.4 In this exercise, we show how the rules of summability calculus for simple finite sums are used in performing infinitesimal calculus for composite finite sums. Let $f(n) = \sum_{k=0}^{n-1}(1+k)^{-n}$, which we mentioned earlier in Chap. 1. Prove that:

$$\frac{d}{dn}\sum_{k=0}^{n-1}\frac{1}{(1+k)^n} = -\sum_{k=0}^{n-1}\frac{\log(1+k)}{(1+k)^n} + n\big(\zeta(n+1) - \sum_{k=0}^{n-1}\frac{1}{(1+k)^{n+1}}\big) \qquad (3.2.10)$$

(Hint: consider the chain rule and the case of semi-linear simple finite sums).

Exercise 3.5 Use the results of Sect. 3.2.1 to derive the trapezoid rule in numerical integration. More specifically, explain why for equally spaced points $x_0, x_1, \ldots, x_n$, with $x_0 = a$ and $x_n = b$, the following approximation to the integral:

$$\int_a^b f(t)dt \approx \frac{b-a}{2n}\big(f(x_0) + 2f(x_1) + 2f(x_2) + \cdots + 2f(x_{n-1}) + f(x_n)\big)$$

is more accurate than:

$$\int_a^b f(t)dt \approx \frac{b-a}{n}\big(f(x_0) + f(x_1) + f(x_2) + \cdots + f(x_{n-2}) + f(x_{n-1})\big)$$

Exercise 3.6 Consider the composite finite sum $f(n) = \sum_{k=0}^{n-1}(\frac{k}{n})^n$. Use Theorem 3.1 and the identity $\sum_{r=0}^{\infty}\frac{B_r}{r!} = \frac{1}{e-1}$ to prove that $\lim_{n\to\infty} f(n) = \frac{1}{e-1}$. Compare this with the results derived in [Spi06].

3.3 Summary

In this chapter, we generalized the earlier results of summability calculus on simple finite sums to the case of composite finite sums. This includes performing infinitesimal calculus and deriving asymptotic expansions.

References

[Apo99] T.M. Apostol, An elementary view of Euler's summation formula. Am. Math. Mon. **106**(5), 409–418 (1999)

[Ber85] B.C. Berndt, Sums related to the harmonic series or the inverse tangent function, in *Ramanujan's Notebooks* (Springer, New York, 1985)

[Pen02] D.J. Pengelley, Dances between continuous and discrete: Euler's summation formula, in *Proceedings of Euler 2K+2 Conference* (2002)

[Rob55] H. Robbins, A remark on Stirling's formula. Am. Math. Mon. **62**, 26–29 (1955)

[SW] J. Sondow, E. Weisstein, Walli's formula. http://mathworld.wolfram.com/WallisFormula. html

[Spi06] M.Z. Spivey, The Euler-Maclaurin formula and sums of powers. Math. Mag. **79**(1), 61–65 (2006)

[Wei] E. Weisstein, Repeated integral. http://mathworld.wolfram.com/RepeatedIntegral.html

Chapter 4
Analytic Summability Theory

Abstract The theory of summability of divergent series is a major branch of mathematical analysis that has found important applications in engineering and science. It addresses methods of assigning natural values to divergent sums, whose prototypical examples include the Abel summation method, the Cesaro means, and the Borel summability method. As will be established in subsequent chapters, the theory of summability of divergent series is intimately connected to the theory of fractional finite sums. In this chapter, we introduce a generalized definition of series as well as a new summability method for computing the value of series according to such a definition. We show that the proposed summability method is both regular and linear, and that it arises quite naturally in the study of local polynomial approximations of analytic functions. The materials presented in this chapter will be foundational to all subsequent chapters.

Leonhard Euler is often quoted saying: "*To those who ask what the infinitely small quantity in mathematics is, we answer that it is actually zero.*" This view of the infinitesimal appears everywhere in calculus, including in central concepts such as derivatives and integrals. For example, the Cauchy definition of a function's derivative is given by $f'(x) = \lim_{h \to 0} \big(f(x+h) - f(x)\big)/h$, which implicitly assumes that the error term $O(h)$ is zero because it is infinitesimal at the limit $h \to 0$.

Clearly, equating infinitesimal quantities with zero is valid if the number of such quantities remains finite. However, it turns out that many important results in infinitesimal calculus are derived by repeatedly applying a particular process *infinitely* many times. For example, classical proofs of the Taylor series expansion and the Euler-Maclaurin summation formula proceed by repeating the integration by parts indefinitely (see for instance [Šik90, Apo99, Lam01]). In such cases, if integration by parts introduces an infinitesimally small error, then a repeated application of such a process infinitely many times can lead to incorrect results.

© Springer International Publishing AG 2018
I. M. Alabdulmohsin, *Summability Calculus*,
https://doi.org/10.1007/978-3-319-74648-7_4

In fact, mathematicians are very familiar with this phenomenon; it is one reason why the algebraic derivation of the Taylor expansion fails to warn against series divergence!

If infinitesimal errors can create divergent series, is there a systematic way of correcting them? Luckily, the answer is yes, and this can often be achieved using any of the prominent summability methods, such as the Abel summability method, the Euler summation method, the Cesàro means, the Mittag-Leffler summability method, and the Lindelöf summability method. In fact, we will also introduce a new summability method in this chapter, denoted χ, which is more convenient than others in practice.

The basic idea behind *summability* methods is to extend the *definition* of series well beyond the classical conditions of convergence. For example, one simple summability method is to define the value of a series by the limiting average of every pair of successive values. More precisely, we extend the definition of series to the following definition:

$$\sum_{k=0}^{\infty} g(k) \triangleq \lim_{n \to \infty} \left\{ \frac{\sum_{k=0}^{n-1} g(k) + \sum_{k=0}^{n} g(k)}{2} \right\} \qquad (4.0.1)$$

This definition might appear arbitrary at a first sight. However, it is, in fact, a "natural" definition of series for many reasons. First, suppose that the series $\sum_{k=0}^{\infty} g(k)$ exists in the classical sense of the word. Then, it is straightforward to see that Eq. 4.0.1 will correctly recover the true value of the series. Such a desirable property is called *regularity*. Second, the definition in Eq. 4.0.1 is *linear* and *stable*.[1] All of these properties suggest that the method in Eq. 4.0.1 can be thought of as a natural extension to the definition of series. If we apply it to the Grandi series, which oscillates indefinitely between 0 and 1, we obtain the familiar result $1 - 1 + 1 - 1 + \cdots = \frac{1}{2}$. The definition in Eq. 4.0.1 was proposed by Hutton in 1812 [Har49]. We will discuss its generalization later.

Of course, there exist many summability methods in the literature; some are weak while others are strong. In this chapter, we will adopt a more "abstract" definition of series, denoted $\mathfrak{T}$. As will be explained in details later, $\mathfrak{T}$ does not immediately provide us with a recipe for *computing* the values of divergent series. However, we will show that nearly all of the summability methods used in practice are consistent with $\mathfrak{T}$, and, hence, can be used to compute such values. The key advantage of dealing with $\mathfrak{T}$ directly, however, is that it simplifies analysis.

---

[1] A summability method is called:

1. *Regular*: If it agrees with ordinary summation whenever a series converges in the classical sense of the word.
2. *Linear*: If it agrees with the identity $\sum_{k=0}^{\infty} g(k) + \lambda h(k) = \sum_{k=0}^{\infty} g(k) + \lambda \sum_{k=0}^{\infty} h(k)$.
3. *Stable*: If it agrees with the identity $\sum_{k=0}^{\infty} g(k) = g(0) + \sum_{k=0}^{\infty} g(k+1)$.

Consider the following example:

$$\sum_{k=0}^{\infty} (-1)^{1+k}\sqrt{1+k} = \sqrt{1} - \sqrt{2} + \sqrt{3} - \sqrt{4} + \cdots$$

It will become clear later in this chapter that this series has a well-defined value under $\mathfrak{T}$. In order to compute this value, we may rely on summability methods, such as the Abel summability method or χ, which are shown to be consistent with $\mathfrak{T}$. Using either approach gives a value of about 0.3802 for this divergent series.

As will be repeatedly shown in the sequel, the formal method of summation $\mathfrak{T}$ is not a mere artificial construct. Using this generalized definition of series, we will derive the analog of the Euler-Maclaurin summation formula for oscillating sums in Chap. 5, which will allow us to perform many remarkable deeds with ease. In Chap. 7, where we extend the rules of summability calculus to arbitrary discrete functions, this generalized definition of series will be used to prove the Shannon-Nyquist sampling theorem. Indeed, it will be repeatedly demonstrated that the summability of divergent series is a fundamental tool in summability calculus.

4.1 The Definition $\mathfrak{T}$

4.1.1 The $\mathfrak{T}$ Definition of Series

In this section, we present the generalized definition of series, which we denote by $\mathfrak{T}$. We will establish conditions for which popular summability methods can be used to *compute* the $\mathfrak{T}$ value of divergent series, but it is important to keep in mind that the definition $\mathfrak{T}$ is not restricted to any particular method of computation.

Definition 4.1 (The $\mathfrak{T}$ Definition of Series) Given a series $\sum_{k=0}^{\infty} g(k)$, define the function:

$$h(z) = \sum_{k=0}^{\infty} g(k) z^k \qquad (4.1.1)$$

In other words, $h(z)$ is the function whose Taylor series expansion around the origin is given by Eq. 4.1.1. If $h(z)$ is analytic in the domain $z \in [0, 1]$, then the $\mathfrak{T}$ value of the series $\sum_{k=0}^{\infty} g(k)$ is defined by $h(1)$.

Note that the formal method of summation $\mathfrak{T}$ not only requires that the function $h(z)$ be analytic at $z = 0$ and $z = 1$, but it also requires that the function be analytic throughout the line segment $[0, 1]$. In other words, we require that the function $h(z)$ be regular at the origin and that the point $z = 1$ falls within the *Mittag-Leffler* star of the function $h(z)$. The Mittag-Leffler star of a function $h(z)$ is the set of all points

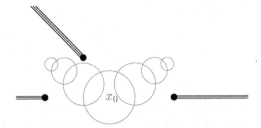

Fig. 4.1 An illustration to the Mittag-Leffler star of a function. Here, the series expansion is taken around x_0 and singularities are identified with solid circles. The rays beyond the singularity points are outside the Mittag-Leffler star of the function. All other points can be reached via a sequence of analytic discs

$z \in \mathbb{C}$ such that the line segment $[0, z]$ does not pass through a singularity point of $h(z)$ [Har49]. This is illustrated in Fig. 4.1.

Looking into Fig. 4.1, we immediately note that the $\mathfrak{T}$ value of series can be computed using *rigidity* of analytic functions. That is, it can be computed by evaluating higher order derivatives $h^{(r)}(z_j)$ at a finite sequence of points $\{z_0, z_1, \ldots, z_{n-1}, z_n\}$ such that z_j falls within the radius of convergence of the Taylor series expansion at z_{j-1}, and $z_0 = 0$ and $z_n = 1$. Because $h(z)$ is analytic at each of those points, this method implies that the $\mathfrak{T}$ value of series is *unique*.

In addition, we have the following alternative method of computing the $\mathfrak{T}$ value of series.

Proposition 4.1 *If a series $\sum_{k=0}^{\infty} g(k)$ has a value in $\mathfrak{T}$ given by $V \in \mathbb{C}$, then V can be computed using either the Mittag-Leffler summability method or the Lindelöf summability method.*

Proof Because the two methods correctly evaluate any Taylor series expansion in the Mittag-Leffler star of the respective function [Har49].[2]   □

Remark 4.1 Note that Proposition 4.1 implies that $\mathfrak{T}$ is consistent with, but is weaker than, both the Mittag-Leffler summability method and the Lindelöf summability method. However, it is important to decouple the generalized definition of series $\mathfrak{T}$ from the method of computation because we are not concerned about any peculiar properties of summability methods. Whether the Mittag-Leffler summability method or the Lindelöf summability method is used is irrelevant to us.

Next, we have the following immediate result.

Proposition 4.2 *The $\mathfrak{T}$ definition of series is regular, linear, and stable. Moreover, if two series $\sum_{k=0}^{\infty} g_1(k)$ and $\sum_{k=0}^{\infty} g_2(k)$ are both defined in $\mathfrak{T}$ by the values V_1 and V_2 respectively, then the $\mathfrak{T}$ value of their Cauchy product is given by the product $V_1 \cdot V_2$.*

---

[2]The reader may consult the Glossary section at the back of this book for brief definitions of these summability methods.

Proof This follows from Definition 4.1 and well-known properties of Taylor series expansions. □

The motivation behind the use of $\mathfrak{T}$ as a generalized definition of series is threefold. First, it is in line with Euler's original reasoning that a mathematical expression should assume the value of the algebraic process that led to it, which is the point of view discussed earlier in Chap. 1. In principle, we opt to select a *natural* assignment of divergent series by using $\mathfrak{T}$ as a definition.

Second, as discussed earlier, we know by *rigidity* of analytic functions that $\mathfrak{T}$ is well-defined. However, in order to guarantee consistency, the conditions of $\mathfrak{T}$ must be valid. For example, the Taylor series expansion of the function $e^{-x^{-2}}$ is the zero function, which might suggest that the sum $(0+0+0+\ldots)$ can be assigned the value e^{-1}! However, because the function is not analytic at the origin, this assignment is invalid. Similarly, ambiguity can arise if a discontinuity exists in the line segment $[0, 1]$. For example, the series $1 + \frac{2^2}{2} + \frac{2^3}{3} + \ldots$ arises out of the series expansion of $\log(1 + x)$ evaluated at $x = -2$. However, both $i\pi$ and $-i\pi$ are equally valid substitutes for $\log(-1)$ and there is no reason to prefer one over the other. Here, we know that the conditions of $\mathfrak{T}$ are not satisfied, so the series is simply undefined in $\mathfrak{T}$. These phenomena never occur if the conditions of Definition 4.1 are satisfied.

Third, because the $\mathfrak{T}$ definition of series is regular, linear, and stable, then arithmetic operations remain consistent even if such operations ultimately lead to convergent series. It might seem implausible at first, but it even occasionally *simplifies* analysis to express convergent series as a linear combination of divergent series, where the latter are defined under $\mathfrak{T}$! For example, we will derive an exact value of the convergent series:

$$\sum_{k=1}^{\infty} (-1)^k (H_k + \gamma - \log k)$$

later in Chap. 5, by deriving analytic values to each of the divergent series $\sum_{k=1}^{\infty} (-1)^k H_k$, $\sum_{k=1}^{\infty} (-1)^k \gamma$, and $\sum_{k=1}^{\infty} (-1)^k \log k$.

4.1.2 The $\mathfrak{T}$ Definition of Sequence Limit

The generalized definition of series yields an immediate generalized definition of limits that is interpreted for the space of *sequences* $S = (s_0, s_1, s_2, \ldots)$. Here, whereas $\mathfrak{T}$ is a generalized definition that assigns values to series, we can reformulate it such that it is interpreted as a method of assigning limits to infinite *sequences*. Clearly, both are essentially equivalent since assigning a value to a series $\sum_{k=0}^{\infty} g(k)$ can be thought of as a method of assigning a limit to the infinite sequence $(\sum_{k=0}^{0} g(k), \sum_{k=0}^{1} g(k), \sum_{k=0}^{2} g(k), \ldots)$.

Definition 4.2 (The $\mathfrak{T}$ Sequence Limit)   Given an infinite sequence $S = (s_0, s_1, s_2, \ldots)$, then the $\mathfrak{T}$ sequence limit is defined by the $\mathfrak{T}$-value of the series $s_0 + \sum_{k=0}^{\infty} \Delta s_k$, where $\Delta s_k = s_{k+1} - s_k$ is the forward difference operator.

The $\mathfrak{T}$ definition of sequence limits appears, at first, to be trivial but it does indeed have interesting consequences. To be more specific, if a finite sum $\sum_{k=0}^{n-1} g(k)$ can be expressed in a closed-form expression $f_G(n)$, then the $\mathfrak{T}$ value of the series $\sum_{k=0}^{\infty} g(k)$ can be obtained by taking the $\mathfrak{T}$ sequence limit of $f_G(n)$ as $n \to \infty$ (see, for example, Exercise 4.2).

The most important special case of $\mathfrak{T}$ sequence limits is stated in the following lemma.

Proposition 4.3 *If the value of a series $\sum_{k=0}^{\infty} g(k)$ exists in $\mathfrak{T}$, then the $\mathfrak{T}$ sequence limit of $(g(k))_{k=0,1,2,\ldots}$ is zero.*

Proof By stability of $\mathfrak{T}$, we have:

$$\sum_{k=0}^{\infty} g(k) = g(0) + \sum_{k=0}^{\infty} g(k+1) \qquad (4.1.2)$$

Because the generalized definition $\mathfrak{T}$ is linear, we have:

$$g(0) + \sum_{k=0}^{\infty} g(1+k) - \sum_{k=0}^{\infty} g(k) = 0 \qquad \Rightarrow g(0) + \sum_{k=0}^{\infty} \Delta g(k) = 0 \qquad (4.1.3)$$

However, the last equation is exactly the $\mathfrak{T}$ definition of the sequence limit of $g(k)$ so the statement of the lemma holds.   □

Proposition 4.3 presents a generalization to convergent series. Traditionally, we know that a series $\sum_{k=0}^{\infty} g(k)$ converges only if $g(k) \to 0$ as $k \to \infty$. Proposition 4.3 states that a series $\sum_{k=0}^{\infty} g(k)$ has a definite value $V < \infty$ under the generalized definition $\mathfrak{T}$ only if the $\mathfrak{T}$ sequence limit of $g(k)$ is zero.

Exercise 4.1 Show that if a sequence $s_0, s_1, s_2, \ldots$ is periodic with period T, then the $\mathfrak{T}$ sequence limit of the sequence is its average $\frac{1}{T} \sum_{t=0}^{T} s_t$. Use this to conclude that the $\mathfrak{T}$ sequence limit of $1, 0, 1, 0, 1, 0, \cdots$ is $\frac{1}{2}$, whereas the $\mathfrak{T}$ sequence limit of $1, -1, 1, -1, \ldots$ is zero. (Hint: if $f(x) = \sum_{k=0}^{\infty} a_k x^k$ and a_k is periodic with period p, then $f(x) = \sum_{k=0}^{p-1} a_k x^k + x^p f(x)$.)

Exercise 4.2 Prove by induction that $\sum_{k=0}^{n} (-1)^k = \frac{1}{2} + (-1)^n \frac{1}{2}$. Use this closed-form expression and Exercise 4.1 to argue that since the $\mathfrak{T}$ sequence limit of $(-1)^n$ as $n \to \infty$ is zero, the $\mathfrak{T}$ value of the series $\sum_{k=0}^{\infty} (-1)^k$ is $\frac{1}{2}$.

Exercise 4.3 Use the linearity and stability of $\mathfrak{T}$ to prove that if $\sum_{k=0}^{\infty} x^k$ exists under $\mathfrak{T}$, then its value must be equal to $(1-x)^{-1}$. Show that $\sum_{k=0}^{\infty} x^k$ is not defined under $\mathfrak{T}$ in the ray $[1, \infty)$. (Hint: consider the Mittag-Leffler star of the function $1/(1-x)$)

4.1.3 Relations to Other Summability Methods

Having a generalized definition of series and limits is crucial, but coming up with a method of *computing* their values is also equally important. Luckily, there exist many methods in the literature for computing the $\mathfrak{T}$ value of series, two of which were already listed in Proposition 4.1.

The following lemma presents other methods for calculating the $\mathfrak{T}$ value of series.

Proposition 4.4 *The $\mathfrak{T}$ definition of series is consistent with, but is more powerful than, the Abel summability method and all Nörlund means including the Cesàro summability method.*[3]

Proof Given the series $\sum_{k=0}^{\infty} g(k)$, define $h(z) = \sum_{k=0}^{\infty} g(k) z^k$. If $\lim_{z \to 1^-} h(z)$ exists, then $h(z)$ is analytic in the domain $[0, 1)$. By Abel's theorem on power series, we have $\lim_{z \to 1^-} h(z) = h(1)$, if $h(z)$ is analytic at $z = 1$. However, assigning a value to the series $\sum_{k=0}^{\infty} g(k)$ using $h(1)$ is the definition of the Abel summability method. If such a limit exists, i.e. if a series is Abel summable, then its value coincides with the $\mathfrak{T}$ definition of the series $\sum_{k=0}^{\infty} g(k)$.

Finally, because the Abel summability method is consistent with, and is more powerful than, all Nörlund means [Har49], the statement of the lemma follows for all Nörlund means, including the Cesàro summability method. □

Proposition 4.4 shows that many methods can be used to compute the $\mathfrak{T}$ value of series. For example, one of the simplest of all Nörlund means is to look into the average of q consecutive partial sums and to determine if these q-averages converge. In the latter case, we have ordinary summation when $q = 1$. If, on the other hand, $q = 2$, we obtain the earlier Hutton summability method in Eq. 4.0.1. Moreover, the case of $q = \infty$, when properly interpreted, leads to the Cesàro summability method. Most importantly, Proposition 4.4 states that if any of those methods yields a value $V \in \mathbb{C}$, then V is the $\mathfrak{T}$ value of the series.

In addition to evaluating series, we can often determine if a series is *not* defined under $\mathfrak{T}$ as the following lemma shows.

---

[3]The Nörlund mean is a method of assigning limits to infinite sequences. Here, suppose p_j is a sequence of positive terms that satisfies $\frac{p_n}{\sum_{k=0}^{n} p_k} \to 0$. Then, the Nörlund mean of a sequence $(s_0, s_1, \dots)$ is given by $\lim_{n \to \infty} \frac{p_n s_0 + p_{n-1} s_1 + \dots + p_0 s_n}{\sum_{k=0}^{n} p_k}$. The limit of an infinite sequence $(s_0, s_1, \dots)$ is defined by its Nörlund mean. Therefore, the Nörlund mean interprets the limit of s_n as $n \to \infty$ using an *averaging* method.

Proposition 4.5 *The following three conditions:*

1. $g(k) \in \mathbb{R}$ *for all* $k \in \mathbb{N}$
2. *The series* $\sum_{k=0}^{\infty} g(k)$ *diverges*
3. *The series* $\sum_{k=0}^{\infty} g(k)$ *has a definite value under* $\mathfrak{T}$.

imply that the terms of the sequence $(g(k))_{k=0,1,2,\dots}$ *must oscillate in sign infinitely many times. In other words, the* $\mathfrak{T}$ *definition of infinite series is* totally regular.[4]

Proof By Proposition 4.1, if $\sum_{k=0}^{\infty} g(k)$ is defined under $\mathfrak{T}$ to some value $V \in \mathbb{C}$, then it must be summable using the Lindelöf summability method to the same value V. However, the Lindelöf summability method assigns to a series the value:

$$\sum_{k=0}^{\infty} g(k) \triangleq \lim_{\delta \to 0} \sum_{k=0}^{\infty} k^{-\delta k} g(k)$$

Because $\lim_{\delta \to 0} k^{-\delta k} = 1$ and $k^{-\delta k} > 0$, the Lindelöf summability method cannot sum a divergent series whose terms $g(k)$ do not oscillate in sign infinitely many times. So, the statement of the lemma follows.   □

Remark 4.2 Proposition 4.5 can be extended to complex-valued functions by considering the real and imaginary parts of $g(k)$ separately.

Proposition 4.5 does not necessarily limit the applicability of the generalized definition $\mathfrak{T}$. For example, whereas $\zeta(s)$ is not directly defined under $\mathfrak{T}$ for $s < 1$, it can be *defined* using Eq. 4.1.4, which is known to hold in the half-plane $\mathfrak{R}(s) < -1$. Therefore, the definition in Eq. 4.1.4 is valid by analytic continuation.

$$\sum_{k=0}^{\infty} (1+k)^s = \frac{1}{1 - 2^{1+s}} \sum_{k=0}^{\infty} (-1)^k (1+k)^s \tag{4.1.4}$$

Now, the right-hand side is well-defined under $\mathfrak{T}$. In fact, we can immediately derive a closed-form expression for it. First, define $N_s(x)$ as:

$$N_s(x) = \sum_{k=1}^{\infty} (-1)^k k^s x^k \tag{4.1.5}$$

Then, $N_s(x)$ satisfies the following recurrence relationship:

$$N_s(x) = x N'_{s-1}(x) \tag{4.1.6}$$

Differentiating both sides of Eq. 4.1.6 with respect to x yields:

$$N'_s(x) = N'_{s-1}(x) + x N^{(2)}_{s-1}(x) \tag{4.1.7}$$

---

[4]Consult the classic book "Divergent Series" by Hardy [Har49] for a definition of this term.

Therefore, upon a repeated application of Eqs. 4.1.6 and 4.1.7, we have:

$$N_s(x) = \sum_{k=0}^{s} S(s,k) \, x^k \, N_0^{(k)}(x) \tag{4.1.8}$$

Here, $S(s,k)$ are the Stirling Numbers of the Second Kind. A simple proof for Eq. 4.1.8 follows by induction upon using the characteristic property of the Stirling Numbers of the Second Kind: $S(n+1,k) = k\,S(n,k) + S(n,k-1)$. Now, knowing that the series expansion of $N_0(x)$ around $x = 1$ is given by Eq. 4.1.9, and upon using Definition 4.1, we arrive at Eq. 4.1.10.

$$N_0(x) = -\frac{x}{1+x} = -\frac{1}{2} - \frac{x-1}{4} + \frac{(x-1)^2}{8} - \frac{(x-1)^3}{16} + \cdots \tag{4.1.9}$$

$$N_s(1) = \sum_{k=1}^{\infty}(-1)^k k^s = \sum_{k=0}^{s}(-1)^k \, S(s,k) \, \frac{k!}{2^{k+1}} \tag{4.1.10}$$

Note here that we have appealed to the original statement of $\mathfrak{T}$ to define such a divergent series. Therefore, we have:

$$\sum_{k=1}^{\infty} k^s = -\frac{1}{1-2^{1+s}} \sum_{k=0}^{s}(-1)^{k+1} \, S(s,k) \, \frac{k!}{2^{k+1}} \tag{4.1.11}$$

Using the identity $B_{1+s} = -(1+s)\zeta(-s)$ for $s \geq 1$, where B_k is the kth Bernoulli number, we have the following closed-form expression for Bernoulli numbers:

$$B_s = \frac{s}{1-2^s} \sum_{k=0}^{s-1} \frac{1}{2^{k+1}} \sum_{j=0}^{k} (-1)^j \binom{k}{j} j^{s-1}, \qquad s \geq 2 \tag{4.1.12}$$

Remark 4.3 Not all oscillating series are summable even if their terms are bounded. For instance, consider the *regular paper-folding* sequence $(1,1,-1,1,1,-1,-1,\ldots)$ (see OEIS A014577 [SWa] and [All15]). Even though the terms of this sequence are in the set $\{-1,+1\}$, the series $\sum_{k=0}^{\infty} s_k$ does not appear to be summable to any value. This fact becomes more remarkable if we note that the regular paper-folding sequence is the limit of a process in which *every* sequence generated is summed exactly to one, yet the limit itself is not summable to any particular value using common summability methods!

Next, we show that the Euler summation method almost always agrees with the $\mathfrak{T}$ definition of series.

Proposition 4.6 (The Euler Sum)   *For an alternating series $\sum_{k=0}^{\infty}(-1)^k g(k)$, define its Euler sum by the following value if it exists:*

$$\sum_{k=0}^{\infty}(-1)^k g(k) \triangleq \sum_{k=0}^{\infty}\frac{(-1)^k}{2^{k+1}}\Delta^k g,$$

where $\Delta g = g(1) - g(0)$, $\Delta^2 g = g(2) - 2g(1) + g(0)$, and so on. Then, the Euler sum is consistent with the $\mathfrak{T}$ definition of series if $\lim_{m\to\infty}\{2^{-m}\sum_{k=0}^{\infty}(-1)^k \Delta^m g(k)\} = 0$.

Proof If the series $\sum_{k=0}^{\infty}g(k)$ exists in $\mathfrak{T}$, then it must satisfy the linearity and stability properties. Thus, we always have:

$$\sum_{k=0}^{\infty}(-1)^k g(k) = \sum_{k=0}^{\infty}(-1)^k g(k+1) - \sum_{k=0}^{\infty}(-1)^k \Delta g(k) \tag{4.1.13}$$

Because the $\mathfrak{T}$ definition is stable, we rewrite the last equation into:

$$\sum_{k=0}^{\infty}(-1)^k g(k) = g(0) - \sum_{k=0}^{\infty}(-1)^k g(k) - \sum_{k=0}^{\infty}(-1)^k \Delta g(k) \tag{4.1.14}$$

Rearranging the terms yields:

$$\sum_{k=0}^{\infty}(-1)^k g(k) = \frac{g(0)}{2} - \frac{1}{2}\sum_{k=0}^{\infty}(-1)^k \Delta g(k) \tag{4.1.15}$$

Here, both divergent series are interpreted using the definition $\mathfrak{T}$. By a repeated application of Eq. 4.1.15, we have:

$$\sum_{k=0}^{\infty}(-1)^k g(k) = \sum_{p=0}^{m}\frac{(-1)^p}{2^{p+1}}\Delta^p g + \frac{(-1)^{m+1}}{2^{m+1}}\sum_{k=0}^{\infty}(-1)^k \Delta^{m+1} g(k) \tag{4.1.16}$$

Last equation gives the error term of using the Euler summability method, which is $\frac{(-1)^{m+1}}{2^{m+1}}\sum_{k=0}^{\infty}(-1)^k \Delta^{m+1} g(k)$. Therefore, if the error term goes to zero and the sum $\sum_{p=0}^{\infty}\frac{(-1)^p}{2^{p+1}}\Delta^p g$ is convergent, then the statement of the lemma follows. $\square$

Proposition 4.6 states that the Euler summation method can indeed be often used to compute the $\mathfrak{T}$ value of divergent series. For instance, if we return to the Riemann zeta function, then the condition of Proposition 4.6 holds so we always have:

$$\sum_{k=1}^{\infty}(-1)^{k+1} k^s = \sum_{k=0}^{\infty}\frac{1}{2^{k+1}}\sum_{j=0}^{k}(-1)^j \binom{k}{j}(j+1)^s \tag{4.1.17}$$

Using Eq. 4.1.17, we deduce a *globally* valid expression for the Riemann zeta function given in Eq. 4.1.18. The expression in Eq. 4.1.18 was proved by Helmut Hasse in 1930 and rediscovered by Sondow in [Son94, SWb]. We will provide an alternative globally convergent expression for the Riemann zeta function later in Chap. 7.

$$\zeta(s) = \frac{1}{1 - 2^{1-s}} \sum_{k=0}^{\infty} \frac{1}{2^{k+1}} \sum_{j=0}^{k} (-1)^j \binom{k}{j} \frac{1}{(j+1)^s} \qquad (4.1.18)$$

Consequently, we indeed have a rich collection of methods to compute the $\mathfrak{T}$ value of series. Unfortunately, most of these methods have serious limitations. For instance, all Nörlund means are weak, i.e. are limited in their ability to sum divergent series, and the Abel summability method alone outperforms them all. However, the Abel summability method is often insufficient. For instance, it cannot evaluate any oscillating series whose absolute terms grow exponentially fast, such as the Taylor series expansion of the logarithmic function and the sum of Bernoulli numbers. On the other hand, the Mittag-Leffler summability method and Lindelöf summation method are both powerful but they are computationally expensive. They are often extremely slow in convergence and require high-precision arithmetics.

To circumvent such limitations, we will introduce a new summability method χ. The method χ is easy to implement in practice and can converge reasonably fast. In addition, it is also powerful and can sum a large number of divergent series. In fact, nearly all of the divergent series presented in this monograph are themselves χ summable.

4.2 The Summability Method χ

In this section, we present a new summability method for Taylor series expansions that is consistent with the $\mathfrak{T}$ definition of series. The method is simple to implement in practice, and its error term is $O(\frac{1}{n})$.

4.2.1 Preliminaries

Definition 4.3 (The χ-Sum) Let $(g(k))_{k=0,1,2,\ldots} \in \mathbb{C}^{\infty}$ be an infinite sequence of complex numbers. Then, the sequence $(g(k))_{k=0,1,2,\ldots}$ will be called χ-summable if the following limit exists:

$$V = \lim_{n \to \infty} \left\{ \sum_{k=0}^{n} \chi_n(k)\, g(k) \right\}, \qquad (4.2.1)$$

(continued)

Definition 4.3 (continued)

where $\chi_n(k) = \prod_{j=1}^{k}\left(1 - \frac{j-1}{n}\right)$ and $\chi_n(0) = 1$. In addition, the limit V, if it exists, will be called the χ-sum of the infinite sequence $(g(k))_{k=0,1,2,\dots}$.

Definition 4.4 (The χ-Limit) Let $(s_k)_{k=0,1,2,\dots} \in \mathbb{C}^\infty$ be an infinite sequence of complex numbers. Define Then, the χ-limit of the sequence $(s_k)_{k=1,2,\dots}$ is defined by the following limit if it exists:

$$\lim_{n\to\infty}\left\{\frac{\sum_{k=0}^{n} p_n(k)\, s_k}{\sum_{k=0}^{n} p_n(k)}\right\}, \tag{4.2.2}$$

where $p_n(k) = k\,\chi_n(k)$.

Later, we will show that both Definitions 4.3 and 4.4 are equivalent to each other. That is, the χ-limit of the sequence of partial sums:

$$\left(\sum_{k=0}^{0} g(k), \sum_{k=0}^{1} g(k), \sum_{k=0}^{2} g(k), \dots\right)$$

is equal to the χ-sum of the infinite sequence $(g(k))_{k=1,2,\dots}$. This shows that the summability method in Definition 4.3, henceforth referred to as χ, is indeed an averaging method, and allows us to prove additional useful properties. Before we do that, we, first, show how the summability method χ arises quite naturally in local polynomial approximations and how it relates to the formal definition of series $\mathfrak{T}$.

4.2.2   Summability and the Taylor Series Approximation

As a starting point, suppose we have a function $f(x)$ that is n-times differentiable at a particular point x_0 and let $h \ll 1$ be a chosen small step size such that we wish to approximate the value of the function $f(x_0 + nh)$ for an arbitrary value of n using *solely* the local information about the behavior of f at x_0. Using the notation $f_j^{(k)} \doteq f^{(k)}(x_0 + jh)$ to denote the k-th derivative of f at the point $x_0 + jh$, we know that the following approximation holds when $h \ll 1$:

$$f_j^{(k)} \approx f_{j-1}^{(k)} + f_{j-1}^{(k+1)} h \tag{4.2.3}$$

In particular, we obtain the following approximation that can be held with an arbitrary accuracy for a sufficiently small step size h:

$$f_n \approx f_{n-1} + f_{n-1}^{(1)} h \qquad (4.2.4)$$

However, the approximation in Eq. 4.2.3 can now be applied to the right-hand side of Eq. 4.2.4. In general, we can show by induction that a repeated application of Eq. 4.2.4 yields the following general formula:

$$f_n \approx \sum_{k=0}^{n} \binom{n}{k} f_0^{(k)} h^k \qquad (4.2.5)$$

To prove that Eq. 4.2.5 holds, we first note that a base case is established for $n = 1$ in Eq. 4.2.4. Suppose that it holds for $n < m$, we will show that this inductive hypothesis implies that Eq. 4.2.5 also holds for $n = m$. First, we note that if Eq. 4.2.5 holds for $n < m$, then we have:

$$f_m \approx f_{m-1} + f_{m-1}^{(1)} h \approx \sum_{k=0}^{m-1} \binom{m-1}{k} f_0^{(k)} h^k + \sum_{k=0}^{m-1} \binom{m-1}{k} f_0^{(k+1)} h^{k+1} \qquad (4.2.6)$$

In Eq. 4.2.6, the second substitution for $f_{m-1}^{(1)}$ follows from the same inductive hypothesis because $f_{m-1}^{(1)}$ is simply another function that can be approximated using the same inductive hypothesis. Equation 4.2.6 can, in turn, be rewritten as:

$$f_m \approx \sum_{k=0}^{m-1} \left[\binom{m-1}{k} + \binom{m-1}{k-1} \right] f_0^{(k)} h^k + f_0^{(m)} h^m \qquad (4.2.7)$$

Upon using the well-known recurrence relation for binomial coefficients, i.e. Pascal's rule, we obtain Eq. 4.2.5, which is precisely what is needed in order to complete the proof by induction of that approximation.

In addition, Eq. 4.2.5 can be rewritten as given in Eq. 4.2.8 below using the substitution $x = x_0 + nh$.

$$f(x) \approx \sum_{k=0}^{n} \binom{n}{k} f^{(k)}(x_0) \frac{(x - x_0)^k}{n^k} \qquad (4.2.8)$$

Next, the binomial coefficient can be expanded, which yields:

$$f(x) \approx \sum_{k=0}^{n} \chi_n(k) \frac{f^{(k)}(x_0)}{k!} (x - x_0)^k \qquad (4.2.9)$$

Since $x = x_0 + nh$, it follows that increasing n while holding x fixed is equivalent to choosing a smaller step size h. Because the entire proof is based solely on the linear approximation recurrence given in Eq. 4.2.3, which is an asymptotic relation as $h \to 0$, the approximation error in Eq. 4.2.9 will typically vanish as $n \to \infty$. Intuitively, this holds because the summability method χ uses a sequence of first-order approximations for the function $f(x)$ in the domain $[x_0, x]$, which is similar to the Euler method. However, contrasting the expression in Eq. 4.2.9 with the classical Taylor series expansion for $f(x)$ gives rise to the χ summability method in Definition 4.3.

Remark 4.4 Interestingly, the summability method χ can be stated succinctly using the language of *symbolic methods*. Here, if we let D be the *differential operator*, then Definition 4.3 states that:

$$f(x) = \lim_{n\to\infty} \left(1 + \frac{xD}{n}\right)^n, \qquad (4.2.10)$$

where the expression is interpreted by expanding the right-hand side into a power series of x and D^k is interpreted as $f^{(k)}(0)$. On the other hand, a Taylor series expansion states that $f(x) = e^{xD}$. Of course, both expressions are equivalent if D were an ordinary number, but they differ when D is a functional operator, as illustrated above. We will return to symbolic methods later in Chap. 7.

4.2.3   Properties of the Summability Method χ

Next, we prove that both Definitions 4.3 and 4.4 are equivalent to each other. In particular, the summability method χ can be interpreted as an averaging method on the sequence of partial sums. In addition, we prove that it is both regular and linear.

Proposition 4.7 *Let* $(s_k)_{k=0,1,\dots} \in \mathbb{C}^\infty$ *be a sequence of partial sums* $s_k = \sum_{j=0}^{k} g(j)$ *for some* $(g(k))_{k=0,1,2,\dots} \in \mathbb{C}^\infty$. *Then the χ-limit of* $(s_k)_{k=1,2,\dots}$ *given by Definition 4.4 is equal to the χ-sum of* $(g(k))_{k=1,2,\dots}$ *given by Definition 4.3.*

Proof Before we prove the statement of the proposition, we first prove the following useful fact:

$$\sum_{k=0}^{n} k\,\chi_n(k) = n \qquad (4.2.11)$$

This holds because:

$$\sum_{k=0}^{n} k\,\chi_n(k) = n! \sum_{k=0}^{n} \frac{k}{(n-k)!\,n^k} = \frac{n!}{n^n} \sum_{k=0}^{n} \frac{n-k}{k!} n^k$$

$$= n! \left(\frac{n}{n^n} \sum_{k=0}^{n} \frac{n^k}{k!} - \frac{n}{n^n} \sum_{k=0}^{n} \frac{n^k}{k!} + \frac{1}{(n-1)!} \right)$$

$$= \frac{n!}{(n-1)!} = n$$

Next, we prove the proposition by induction. First, let us write $c_n(k)$ to denote the sequence of terms that satisfy:

$$\frac{p_n(0)s_0 + p_n(1)s_1 + \cdots + p_n(n)s_n}{\sum_{k=0}^{n} p_n(k)} = \sum_{k=0}^{n} c_n(k)\,g(k) \tag{4.2.12}$$

Again, we have $s_j = \sum_{k=0}^{j} g(k)$. Our objective is to prove that $c_n(k) = \chi_n(k)$. To prove this by induction, we first note that $\sum_{k=0}^{n} p_n(k) = n$ by Eq. 4.2.11, and a base case is already established since $c_n(0) = 1 = \chi_n(0)$. Now, we note that:

$$c_n(k) = \frac{p_n(k) + p_n(k+1) + \cdots + p_n(n)}{n} \tag{4.2.13}$$

For the inductive hypothesis, we assume that $c_n(k) = \chi_n(k)$ for $k < m$. To prove that this inductive hypothesis implies that $c_n(m) = \chi_n(m)$, we note by Eq. 4.2.13 that the following identity holds:

$$c_n(m) = c_n(m-1) - \frac{p_n(m-1)}{n} = \left(1 - \frac{m-1}{n}\right)\chi_n(m-1) = \chi_n(m) \tag{4.2.14}$$

Here, we have used the inductive hypothesis and the original definition of $\chi_n(m)$. Therefore, we indeed have:

$$\frac{p_n(0)s_0 + p_n(1)s_1 + \cdots + p_n(n)s_n}{\sum_{k=0}^{n} p_n(k)} = \sum_{k=0}^{n} \chi_n(k)\,g(k), \tag{4.2.15}$$

which proves the statement of the proposition. □

Proposition 4.7 shows that the summability method χ is indeed an averaging method but it is different from the Nörlund and the Cesáro means because the terms of the sequence $p_n(k)$ depend on n. In fact, whereas the Nörlund and Cesáro means are strictly weaker than the Abel summation method [Har49], the summability method χ is strictly stronger than the Abel summability method. Therefore, the summability method χ is not contained in the Nörlund means.

Next, Proposition 4.7 allows us to prove the following important statement.

Corollary 4.1 *The summability method χ is linear and regular.*

Proof To show that the summability method is linear, we note using the basic laws of limits that:

$$\lim_{n\to\infty}\left\{\sum_{j=0}^{n}\chi_n(\alpha_j+\lambda\,\beta_j)\right\}=\lim_{n\to\infty}\left\{\sum_{j=0}^{n}\chi_n(j)\,\alpha_j\right\}+\lambda\lim_{n\to\infty}\left\{\sum_{j=0}^{n}\chi_n(j)\,\beta_j\right\}$$

Hence, $\chi:\mathbb{C}^\infty\to\mathbb{C}$ is a linear operator. To show regularity, we use the Toeplitz-Schur Theorem. The Toeplitz-Schur Theorem states that any matrix summability method $t_n=\sum_{k=0}^{\infty}A_{n,k}s_k$ $(n=0,1,2,\ldots)$ in which $\lim_{n\to\infty}s_n$ is defined by $\lim_{n\to\infty}t_n$ is regular *if and only if* the following three conditions hold [Har49]:

1. $\sum_{k=0}^{\infty}|A_{n,k}|<H$, for all n and some constant H that is independent of n.
2. $\lim_{n\to\infty}A_{n,k}=0$ for each k.
3. $\lim_{n\to\infty}\sum_{k=0}^{\infty}A_{n,k}=1$.

Using Proposition 4.7, we see that the summability method χ is a matrix summability method characterized by $A_{n,k}=p_n(k)/n=k\,\chi_n(k)/n$. Because $A_{n,k}\geq0$ and $\sum_{k=0}^{\infty}A_{n,k}=\sum_{k=0}^{n}A_{n,k}=1$, both conditions 1 and 3 are immediately satisfied. In addition, for each fixed k, $\lim_{n\to\infty}A_{n,k}=0$, thus condition 2 is also satisfied. Therefore, the summability method χ is regular.[5] $\square$

4.2.4 Convergence

In this section, we derive some properties related to the convergence of the summability method χ. As is customary in the literature (cf. the *Borel-Okada* principle [Har49, Kor04]), we analyze the conditions when χ evaluates the Taylor series expansion of the geometric series $\sum_{k=0}^{\infty}x^k$ to the value $(1-x)^{-1}$. We will focus on the case when $x\in\mathbb{R}$, and provide an asymptotic expression to the error term afterward.

4.2.4.1 The Geometric Series Test

We begin with the following elementary result.

Lemma 4.1 *For any $n\geq1$ and all $x\in[0,n]$, we have $\left|(1-\frac{x}{n})^n-e^{-x}\right|\leq\frac{1}{en}$.*

---

[5]In fact, because $A_{n,k}\geq0$, it is also *totally regular* (see [Har49] for a definition of this term).

Proof Define $g_n(x) = (1 - x/n)^n - e^{-x}$ to be the approximation error of using $(1 - x/n)^n$ for the exponential function e^{-x}. Then:

$$g_n(0) = 0$$

$$|g_n(n)| = e^{-n}$$

$$g_n'(x_0) = 0 \wedge x_0 \in [0, n] \quad \Leftrightarrow \quad e^{-x_0} = \left(1 - \frac{x_0}{n}\right)^{n-1}$$

In the last case, we have $|g(x_0)| = \frac{x_0}{n} e^{-x_0} \le \frac{1}{en}$. Because $(e\,n)^{-1} \ge e^{-n} \ge 0$ for all $n \ge 1$, we deduce the statement of the lemma. $\qquad\qquad\square$

Proposition 4.8 *The χ-sum of the infinite sequence $(x^k)_{k=0,1,2,\dots}$ for $x \in \mathbb{R}$ is $(1 - x)^{-1}$ if and only if $-\kappa < x < 1$, where $\kappa \approx 3.5911$ is the solution to $\kappa \, \log \kappa - \kappa = 1$. For values of $x \in \mathbb{R}$ that lie outside the open set $(-\kappa, 1)$, the infinite sequence $(x^k)_{k=0,1,2,\dots}$ is not χ-summable.*

Proof (Case I). First, we consider the case when $x \ge 1$. Because every term in the sequence x^k is non-negative, $\chi_n(k) \ge 0$, and $\lim_{n\to\infty} \chi_n(k) = 1$ for any fixed $k \ge 0$, we have:

$$\lim_{n\to\infty} \left\{ \sum_{k=0}^{\infty} \chi_n(k) \, x^k \right\} = \infty, \qquad \text{if } x \ge 1$$

More generally, if an infinite sequence $(g(k))_{k=0,1,\dots}$ is χ-summable but $\sum_{k=0}^{\infty} g(k)$ does not exist, then the sequence $(g(k))_{k=0,1,\dots}$ must oscillate in sign infinitely many times. This is another way of stating that the summability method is totally-regular [Har49].

(Case II). Second, we consider the case when $-1 < x < 1$. Because the geometric series converges in this domain and the summability method χ is regular (i.e. consistent with ordinary convergence) as proved in Corollary 4.1, we obtain:

$$\lim_{n\to\infty} \left\{ \sum_{k=0}^{\infty} \chi_n(k) \, x^k \right\} = \frac{1}{1-x}, \qquad \text{if } -1 < x < 1$$

(Case III). Finally, we consider the case when $x \le -1$. To simplify notation, we will write $x = -z$, where $z \ge 1$. We have:

$$\sum_{k=0}^{n} \chi_n(k) \, (-z)^k = \sum_{k=0}^{n} \chi_n(k) \frac{(-z)^k}{k!} \int_0^{\infty} t^k \, e^{-t} \, dt = \int_0^{\infty} \sum_{k=0}^{n} \chi_n(k) \frac{(-zt)^k}{k!} \, e^{-t} \, dt$$

$$= \int_0^{\infty} \left(1 - \frac{tz}{n}\right)^n e^{-t} \, dt$$

$$= \int_0^{\frac{n}{z}} \left(1 - \frac{tz}{n}\right)^n e^{-t} \, dt \quad + \quad \int_{\frac{n}{z}}^{\infty} \left(1 - \frac{tz}{n}\right)^n e^{-t} \, dt$$

Now, we consider each integral separately. First, we use the change of variable $w = 1 - \frac{tz}{n}$ to deduce that the second integral is given by:

$$\int_{\frac{n}{z}}^{\infty} (1 - \frac{tz}{n})^n e^{-t} \, dt = \frac{(n!) \, z^n \, e^{\frac{n}{z}}}{n^n}$$

For the first integral, on the other hand, we use Lemma 4.1 to obtain the bound:

$$\int_0^{\frac{n}{z}} \left| (1 - \frac{tz}{n})^n - e^{-tz} \right| e^{-t} \, dt \leq \frac{1}{en} \int_0^{\frac{n}{z}} e^{-t} \, dt$$

$$= \frac{1 - e^{-n/z}}{en} \rightarrow 0 \quad \text{as } n \to \infty,$$

Taking the limit as $n \to \infty$, we deduce that:

$$\lim_{n \to \infty} \int_0^{\frac{n}{z}} (1 - \frac{tz}{n})^n e^{-t} \, dt = \lim_{n \to \infty} \int_0^{\frac{n}{z}} e^{-t(1+z)} \, dt = \frac{1}{1+z} = \frac{1}{1-x}$$

Therefore, in order for the summability method χ to evaluate the geometric series to the value $(1 - x)^{-1}$, we must have:

$$\lim_{n \to \infty} \left\{ \frac{(n!) \, |x|^n \, e^{\frac{n}{|x|}}}{n^n} \right\} = 0$$

Using Stirling's approximation [Rob55]:

$$\frac{(n!) \, |x|^n \, e^{\frac{n}{|x|}}}{n^n} \sim \left(\frac{|x| \, e^{\frac{1}{|x|}}}{e} \right)^n,$$

which goes to zero as $n \to \infty$ only if $|x| < \kappa$. $\square$

4.2.4.2   Asymptotic Analysis

Next, we derive an asymptotic expression to the error term when the summability method χ is applied to a power series. To do this, we begin with the following lemma.

Lemma 4.2 *Let $\sum_{k=0}^{\infty} a_k (x - x_0)^k$ be a power series for some function $f(x)$ that is analytic throughout the domain $[x_0, x]$. Then:*

$$\lim_{n \to \infty} \left\{ \sum_{k=1}^{n} \chi_n(k) \left[f(x) - \sum_{j=0}^{k-1} \frac{f^{(j)}(x_0)}{j!} (x - x_0)^j \right] \right\} = (x - x_0) f'(x), \qquad (4.2.16)$$

provided that $\sum_{k=0}^{n} \chi_n(k) a_k (x - x_0)^k$ *converges uniformly to* $f(x)$ *in the domain* $[x_0, x]$ *as* $n \to \infty$.

Proof It is straightforward to see that if the limit in Eq. 4.2.16 exists, then Eq. 4.2.16 must hold. We will prove this formally. To see this, define g_f to be a functional of $f(x)$ that is given by:

$$g_f = \lim_{n \to \infty} \left\{ \sum_{k=1}^{n} \chi_n(k) \left[f(x) - \sum_{j=0}^{k-1} \frac{f^{(j)}(x_0)}{j!} (x - x_0)^j \right] \right\}$$
(4.2.17)

Now, we differentiate both sides with respect to x, which yields:

$$\frac{d}{dx} g_f = \lim_{n \to \infty} \left\{ \sum_{k=1}^{n} \chi_n(k) \left[\frac{f^{(k)}(x_0)}{(k-1)!} (x - x_0)^{(k-1)} + f'(x) - \sum_{j=0}^{k-1} \frac{f^{(j+1)}(x_0)}{j!} (x - x_0)^j \right] \right\}$$

$$= g_{f'} + \lim_{n \to \infty} \sum_{k=1}^{n} \chi_n(k) \frac{f^{(k)}(x_0)}{(k-1)!} (x - x_0)^{(k-1)}$$

$$= g_{f'} + f'(x)$$

Therefore, we have:

$$\frac{d}{dx} g_f = f'(x) + g_{f'}$$
(4.2.18)

Since $g_f(x_0) = 0$, the solution is given by:

$$g_f(x) = (x - x_0) f'(x)$$
(4.2.19)

The above derivation assumes that the limit exists. To prove that Eq. 4.2.16 holds under the stated conditions, we note that $f(x) - \sum_{j=0}^{k-1} \frac{f^{(j)}(x_0)}{j!} (x - x_0)^j$ is the error term of the Taylor series expansion, which is *exactly* given by:

$$f(x) - \sum_{j=0}^{k-1} \frac{f^{(j)}(x_0)}{j!} (x - x_0)^j = \int_{x_0}^{x} \frac{f^{(k)}(t)}{(k-1)!} (x - t)^{k-1} dt$$
(4.2.20)

Upon using the last expression, we have that:

$$g_f = \lim_{n \to \infty} \sum_{k=1}^{n} \chi_n(k) \int_{x_0}^{x} \frac{f^{(k)}(t)}{(k-1)!} (x - t)^{k-1} dt$$

$$= \int_{x_0}^{x} \lim_{n \to \infty} \sum_{k=1}^{n} \chi_n(k) \frac{f^{(k)}(t)}{(k-1)!} (x - t)^{k-1} dt$$
(4.2.21)

Here, exchanging sums with integrals is justifiable when uniform convergence holds. However:

$$\lim_{n \to \infty} \sum_{k=1}^{n} \chi_n(k) \frac{f^{(k)}(t)}{(k-1)!} (x - t)^{k-1} = f'(x), \quad \text{for all } t \in [x_0, x] \tag{4.2.22}$$

Plugging Eq. 4.2.22 into Eq. 4.2.21 yields the desired result.   □

Theorem 4.1 Let $\hat{f}_n(x) = \sum_{k=0}^{n} \chi_n(k) \frac{f^{(k)}(x_0)}{k!} (x - x_0)^k$. Then, under the stated conditions of Lemma 4.2, the error as $n \to \infty$ is asymptotically given by:

$$f(x) - \hat{f}_n(x) \sim \frac{f''(x)(x - x_0)^2}{2n} \tag{4.2.23}$$

Proof Because the function $f(x)$ is analytic at every point in the domain $[x_0, x]$, define $\epsilon > 0$ to be the distance between the set $[x_0, x]$ and the nearest singularity point of $f(x)$. More formally, let $\mathcal{B}(z, r) \subset \mathbb{C}$ be the open ball of radius r centered at z, and define:

$$\epsilon = \sup \{r \mid \forall z \in [x_0, x] : f \text{ is analytic in } \mathcal{B}(z, r)\}$$

In our construction of the summability method χ in Sect. 4.2.2, we have used the following linear approximations, where $f_j^{(k)}$ is a shorthand for $f^{(k)}(x_0 + jh)$, and h is a small step size:

$$f_j^{(k)} \approx f_{j-1}^{(k)} + f_{j-1}^{(k+1)} h \tag{4.2.24}$$

Because the distance from $x_0 + jh$ to the nearest singularity point of $f(x)$ is at least ϵ, then selecting $h < \epsilon$ or equivalently $n > \frac{x - x_0}{\epsilon}$ implies that the error term of this linear approximation is *exactly* given by Eq. 4.2.25. This follows from the classical result in complex analysis that an analytic function is equal to its Taylor series representation within its radius of convergence, where the radius of convergence is, at least, equal to the distance to the nearest singularity.

$$E_j^k = f_j^{(k)} - f_{j-1}^{(k)} - f_{j-1}^{(k+1)} h = \sum_{m=2}^{\infty} \frac{f_{j-1}^{(k+m)}}{m!} h^m \tag{4.2.25}$$

Since we seek an asymptotic expression as $n \to \infty$, we assume that n is large enough for Eq. 4.2.25 to hold (i.e. $n > (x - x_0)/\epsilon$). Because higher-order derivatives at x_0 are computed exactly, we have $E_0^k = 0$. Now, the linear approximation method was applied recursively in our construction of χ in Sect. 4.2.2. Visually speaking, this repeated process mimics the expansion of a binary pyramid, depicted in Fig. 4.2, whose nodes (j, k) correspond to the linear approximations given by Eq. 4.2.24 and the two children of each node (j, k) are given by $(j - 1, k)$ and $(j - 1, k + 1)$ as

Fig. 4.2 A depiction of the proof by construction of the summability method χ in Sect. 4.2.2

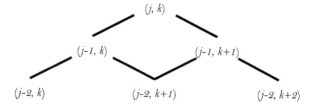

stated in that equation. It follows, therefore, that the number of times the linear approximation in Eq. 4.2.24 is used for a fixed n is equal to the number of paths from the root to the respective node in the binary pyramid, where the root is $(n, 0)$. It is a well-known result that the number of such paths is given by $\binom{n-j}{k}$.

Consequently, the error term of the χ summability method is given by:

$$f(x) - \hat{f}_n(x) = \sum_{j=1}^{n} \binom{n-j}{0} E_j^0 + \sum_{j=1}^{n-1} \binom{n-j}{1} E_j^1 h + \sum_{j=1}^{n-2} \binom{n-j}{2} E_j^2 h^2 \cdots \qquad (4.2.26)$$

Define e_k to be a weighted average of the errors E_j^k that is given by:

$$e_k = \frac{\sum_{j=1}^{n-k} \binom{n-j}{k} E_j^k}{\sum_{j=1}^{n-k} \binom{n-j}{k}} \qquad (4.2.27)$$

Then, we have:

$$f(x) - \hat{f}_n(x) = e_0 \sum_{j=1}^{n} \binom{n-j}{0} + e_1 \sum_{j=1}^{n-1} \binom{n-j}{1} h + e_2 \sum_{j=1}^{n-2} \binom{n-j}{2} h^2 \cdots \qquad (4.2.28)$$

Now, we make use of the identity $\sum_{j=1}^{n-k} \binom{n-j}{k} = \binom{n}{k+1}$, and substitute $h = \frac{x-x_0}{n}$, which yields:

$$f(x) - \hat{f}_n(x) = \frac{n}{x - x_0} \sum_{k=1}^{n} \chi_n(k) \frac{e_{k-1}}{k!} (x - x_0)^k \qquad (4.2.29)$$

Interestingly, the terms $\chi_n(k)$ appear again in the approximation error. Now, define e_k^* by:

$$e_k^* = \frac{e_k}{h^2} = \sum_{j=1}^{n-k} \frac{\binom{n-j}{k}}{\binom{n}{k+1}} \sum_{m=0}^{\infty} \frac{f_{j-1}^{(k+m+2)}}{(m+2)!} h^m \qquad (4.2.30)$$

This yields:

$$f(x) - \hat{f}_n(x) = \frac{x - x_0}{n} \sum_{k=1}^{n} \chi_n(k) \frac{e_{k-1}^*}{k!} (x - x_0)^k \qquad (4.2.31)$$

Next, we examine the weighted average error terms e_k^* at the limit $n \to \infty$. Because we desire an asymptotic expression of $f(x) - f_n(x)$ as $n \to \infty$ up to a first order approximation, we can safely ignore any $o(1)$ errors in the asymptotic expressions for e_{k-1}^* in Eq. 4.2.31.

First, we note that the expression for e_k^* given by Eq. 4.2.30 is *exact* and that the series converges because n is assumed to be large enough such that h is within the radius of convergence of the series (i.e. satisfies $h < \epsilon$). Therefore, it follows that e_k^* is asymptotically given by Eq. 4.2.32, where the expression is asymptotic as $n \to \infty$. This follows from the fact that a Taylor series expansion around a point x_0 is an asymptotic expression as $x \to x_0$ in the Poincaré sense if $f(x)$ is regular at x_0.

$$e_k^* \sim \sum_{j=1}^{n-k} \frac{\binom{n-j}{k}}{\binom{n}{k+1}} \left[\frac{f_{j-1}^{(k+2)}}{2} + o(1) \right] \qquad (4.2.32)$$

Because $p_k(j; n) = \binom{n-j}{k}/\binom{n}{k+1}$ is a normalized discrete probability mass function, and from the earlier discussion, the $o(1)$ term can be safely ignored. This gives us:

$$e_k^* \sim \sum_{j=1}^{n-k} \frac{\binom{n-j}{k}}{\binom{n}{k+1}} \frac{f_{j-1}^{(k+2)}}{2} \qquad (4.2.33)$$

To come up with a more convenient form, we note that the discrete probability mass function $p_k(j; n) = \binom{n-j}{k}/\binom{n}{k+1}$ approaches a probability density function at the limit $n \to \infty$ when $\frac{j}{n}$ is fixed at a constant z. Using Stirling's approximation [Rob55], such a probability density is given by:

$$\rho_k(z) = (1 + k)(1 - z)^k, \qquad 0 \le z \le 1 \qquad (4.2.34)$$

For example, if $k = 0$, the probability density function is uniform as expected.

Upon making the substitution $z = j/n$, the discrete probability mass function $p_k(j; n)$ can be approximated by $\rho_k(z)$ whose approximation error vanishes at the limit $n \to \infty$. Using $\rho_k(z)$, e_k^* is asymptotically given by:

$$e_k^* \sim \frac{1 + k}{2} \int_0^1 (1 - t)^k f^{(k+2)} (x_0 + t(x - x_0)) \, dt \qquad (4.2.35)$$

Doing integration by parts yields the following recurrence identity:

$$e_k^* \sim -\frac{1 + k}{2(x - x_0)} f^{(k+1)}(x_0) + \frac{1 + k}{x - x_0} e_{k-1}^* \qquad (4.2.36)$$

Also, we have by direct evaluation of the integral in Eq. 4.2.34 when $k = 0$:

$$e_0^* \sim \frac{f'(x) - f'(x_0)}{2(x - x_0)} \tag{4.2.37}$$

Combining both Eqs. 4.2.36 and 4.2.37, we obtain the convenient expression:

$$e_k^* \sim \frac{1}{2} \frac{(1+k)!}{(x-x_0)^{k+1}} f'(x) - \frac{1}{2} \frac{(1+k)!}{(x-x_0)^{k+1}} \sum_{m=0}^{k} \frac{f^{(m+1)}(x_0)}{m!} (x - x_0)^m \tag{4.2.38}$$

Plugging Eq. 4.2.38 into Eq. 4.2.31 yields:

$$f(x) - \hat{f}_n(x) \sim \frac{x - x_0}{2n} \sum_{k=1}^{n} \chi_n(k) \left[f'(x) - \sum_{m=0}^{k-1} \frac{f^{(m+1)}(x_0)}{m!} (x - x_0)^m \right] \tag{4.2.39}$$

Using Lemma 4.2, we arrive at the desired result:

$$f(x) - \hat{f}_n(x) \sim \frac{f^{(2)}(x) (x - x_0)^2}{2n} \tag{4.2.40}$$

□

To test the asymptotic expression given by Theorem 4.1, suppose we evaluate the geometric series $f(x) = \sum_{k=0}^{\infty} x^k$ at $x = -2$ using $n = 40$. Then, the error term by a direct application of the summability method χ is 0.0037 (up to four decimal places). The expression in Theorem 4.1 predicts a value of 0.0037, which is indeed accurate up to four decimal places. Similarly, if we apply the summability method χ to the Taylor series expansion $f(x) = \log(1 + x) = \sum_{k=1}^{\infty} \frac{(-1)^{k+1}}{k}$ at $x = 3$ using $n = 30$, then the error term is -0.0091 (up to four decimal places) whereas Theorem 4.1 estimates the error term to be -0.0094.

The asymptotic expression for the error term given in Theorem 4.1 presents an interesting insight. Specifically, we can determine if the summability method converges to a value $f(x)$ from above or from below depending on whether the function f is concave or convex at x. This conclusion becomes easy to explain if we keep in mind that the summability method χ uses first-order linear approximations, in a manner that is quite similar to the Euler method. Thus, at the vicinity of x, first order linear approximation overestimates the value of $f(x)$ if f is concave at x and underestimates it if f is convex.

4.2.5 Examples

In Sect. 4.2.4, we proved that the summability method χ "correctly" evaluates the geometric series $\sum_{k=0}^{\infty} x^k$ in the domain $-\kappa < x < 1$, where $\kappa \approx 3.5911$ is the solution to $\kappa \log \kappa - \kappa = 1$. Because $\kappa > 1$, the summability method is strictly more powerful than the Abel summation method and all the Nörlund and the Cesáro means. In this section, we present two additional examples that demonstrate how the summability method χ indeed assigns "correct" values to divergent series.

We begin with the following example.

4.2.5.1   Euler's Constant

Later in Chap. 5, we will show that the following assignments are valid under $\mathfrak{T}$:

$$\sum_{k=0}^{\infty}(-1)^k H_{k+1} = \frac{\log 2}{2}, \quad \sum_{k=0}^{\infty}(-1)^k \log(1+k) = \log\sqrt{\frac{2}{\pi}}, \quad \sum_{k=0}^{\infty}(-1)^k = \frac{1}{2}$$

$$(4.2.41)$$

These expressions can be easily verified using the summability method χ, which gives the following values when $n = 100$:

$$\sum_{k=0}^{\infty}(-1)^k H_{k+1} \approx 0.3476, \quad \sum_{k=0}^{\infty}(-1)^k \log(1+k) \approx -0.2261, \quad \sum_{k=0}^{\infty}(-1)^k \approx 0.4987$$

The latter values agree with the exact expressions in Eq. 4.2.41 up to two decimal places. A higher accuracy can be achieved using a larger value of n.

However, a famous result, due to Euler, states that the harmonic numbers $H_n = \sum_{k=1}^{n} (1/k)$ are asymptotically related to $\log n$ by the relation $H_n \sim \log n + \gamma$, where $\gamma \approx 0.577$ is the Euler constant. This implies that the alternating series

$$\sum_{k=0}^{\infty}(-1)^k \big(H_{k+1} - \log(k+1) - \gamma\big)$$

converges to some value $V \in \mathbb{R}$. It is relatively simple to derive an *approximate* value of V using, for example, the Euler-Maclaurin summation formula. However, regularity and linearity of the summability method χ can be employed to deduce the *exact* value of the convergent sum. Specifically, we have:

$$V = \sum_{k=0}^{\infty}(-1)^k \big(H_{k+1} - \log(k+1) - \gamma\big) \qquad \text{(by definition)}$$

$$= \lim_{n\to\infty} \sum_{k=0}^{n} \chi_n(k)\,(-1)^k \big(H_{k+1} - \log(1+k) - \gamma\big) \text{ (by regularity)}$$

$$= \left[\lim_{n\to\infty} \sum_{k=0}^{n} \chi_n(k)\,(-1)^k H_{k+1}\right]$$

$$\quad -\left[\lim_{n\to\infty} \sum_{k=0}^{n} \chi_n(k)\,(-1)^k \log(k+1)\right]$$

$$\quad -\gamma\left[\lim_{n\to\infty} \sum_{k=0}^{n} \chi_n(k)\,(-1)^k\right] \qquad \text{(by linearity)}$$

Plugging in the exact values in Eq. 4.2.41, we deduce that:

$$\sum_{k=0}^{\infty}(-1)^k\left(H_{k+1}-\log(k+1)-\gamma\right) = \frac{\log\pi-\gamma}{2}$$

Hence, we used the $\mathfrak{T}$ values (sometimes called *antilimits* in the literature) of divergent series to determine the limit of a convergent series.

4.2.5.2   Bernoulli Numbers

In our second example, we consider the series $f(x) = \sum_{k=0}^{\infty} B_k\,x^k$, where $B_k = (1, -\frac{1}{2}, \frac{1}{6}, 0, \ldots)$ are the Bernoulli numbers. Here, $f(x)$ diverges everywhere except at the origin. Using the Euler-Maclaurin summation formula, it can be shown that $f(x)$ is an asymptotic series to the function $h(x) = \frac{1}{x}\varpi''(1/x)$ as $x \to 0$, where $\varpi(x) = \log(x!)$ is the log-factorial function.[6] Hence, we can indeed contrast the values of the divergent series, when interpreted using the summability method χ, with the values of $h(x)$. However, because B_k increase quite rapidly, the series $\sum_{k=0}^{\infty} B_k\,x^k$ is not χ-summable *per se* but it can be approximated using small values of n, nevertheless.

Table 4.1 lists the values of $\sum_{k=0}^{n} \chi_n(k)\,B_k\,x^k$ for different choices of n and x. As shown in the table, the values are indeed close to the "correct" values, given by $h(x) = \frac{1}{x}\varpi''(1/x)$, despite the fact that the finite sums $\sum_{k=0}^{n} B_k\,x^k$ bear no resemblance with the generating function $h(x)$. Note, for instance, that if $n = 30$, then $\sum_{k=0}^{n} B_k \approx 5.7 \times 10^8$, whereas $\sum_{k=0}^{n} \chi_n(k)B_k \approx 0.6425$, where the latter figure is accurate up to two decimal places. This agrees with Euler's conclusion that $\sum_{k=0}^{\infty} B_k = \zeta(2) - 1 = \pi^2/6 - 1$ [Pen02].

Most importantly, whereas the summability method χ is shown to be useful in this example, even when the series $\sum_{k=0}^{\infty} B_k\,x^k$ diverges quite rapidly, other classical summability methods such as the Abel summation method and the Mittag-Leffler summability method cannot be used in this case. Hence, the summability method χ

Table 4.1 For every choice of x and n, the value in the corresponding cell is that of $\sum_{k=0}^{n} \chi_n(k)\,B_k\,x^k$

	$x = -1$	$x = -0.7$	$x = -0.2$	$x = 0$	$x = 0.2$	$x = 0.7$	$x = 1$
$n = 20$	1.6407	1.4227	1.1063	1.000	0.9063	0.7227	0.6407
$n = 25$	1.6415	1.4233	1.1064	1.000	0.9064	0.7233	0.6415
$n = 30$	1.6425	1.4236	1.1064	1.000	0.9064	0.7236	0.6425
Exact	1.6449	1.4255	1.1066	1.000	0.9066	0.7255	0.6449

The exact value in the last row is that of the generating function $\frac{1}{x}\varpi''(1/x)$, where $\psi(x) = \log(x!)$ is the log-factorial function

---

[6]In MATLAB, the function is given by the command: `1/x*psi(1,1/x+1)` .

can be quite useful in computing the values of divergent series when other methods fail.

Exercise 4.4 Verify that $\sum_{k=0}^{\infty}(-1)^k(1 + k)^s$ is χ-summable to $(1 - 2^{1-s}) \zeta(-s)$, where $\zeta(s)$ is the Riemann zeta function. Use this fact and Proposition 4.8 to argue that:

$$\lim_{n\to\infty}\left\{ \sum_{k=0}^{n}(-1)^k \chi_n(k)(1 + k)^s \right\} = \lim_{n\to\infty}\left\{ n! \sum_{k=0}^{n}(-1)^k \frac{(1 + k)^s}{n^k(n - k)!} \right\}$$

$$= (1 - 2^{1+s}) \zeta(-s)$$

Exercise 4.5 Use the Euler-Maclaurin summation formula to derive the following formal expression for Euler's constant:

$$\gamma = 1 + \sum_{r=1}^{\infty} \frac{B_r}{r}$$

This series diverges rapidly. Show that the χ-summability method with small values of n, e.g. $n = 10, 20, 30$, yields values that agree with the above formal expression.

Exercise 4.6 Show that:

$$\sum_{k=1}^{\infty}(-1)^{k+1}\frac{\sin(kx)}{k} = \frac{x}{2} \qquad\qquad (4.2.42)$$

Verify this result using the χ summability method for various choices of x. Integrate formally both sides of Eq. 4.2.42 and use it to show that $\zeta(2) = \frac{\pi^2}{6}$. (Hint: start with the fact that $\sin(kx)$ can be represented in terms of complex exponential functions and use Exercise 4.3)

4.3 Summary

In this chapter, a generalized definition of series $\mathfrak{T}$ is presented, which is regular, linear, stable, and respects the Cauchy product. To compute the $\mathfrak{T}$ value of series, various methods can be employed, such as the Abel summability method and the Lindelöf summation methods. In addition, a new summability method χ is proposed in this chapter that is simple to implement in practice and is powerful enough to sum nearly all of the divergent series mentioned in this monograph.

In the following chapter, we will use these tools to deduce the analog of the Euler-Maclaurin summation formula for oscillating sums, which will simplify the study of oscillating sums considerably and shed further insights into the subject of divergent series.

References

[All15] J.-P. Allouche, Paperfolding infinite products and the gamma function. J. Number Theory **148**, 95–111 (2015)

[Apo99] T.M. Apostol, An elementary view of Euler's summation formula. Am. Math. Mon. **106**(5), 409–418 (1999)

[Har49] G.H. Hardy, *Divergent Series* (Oxford University Press, New York, 1949)

[Kor04] J. Korevaar, *Tauberian Theory: A Century of Developments* (Springer, Berlin, 2004)

[Lam01] V. Lampret, The Euler-Maclaurin and Taylor formulas: twin, elementary derivations. Math. Mag. **74**(2), 109–122 (2001)

[Pen02] D.J. Pengelley, Dances between continuous and discrete: Euler's summation formula, in *Proceedings of Euler 2K+2 Conference* (2002)

[Rob55] H. Robbins, A remark on Stirling's formula. Am. Math. Mon. **62**, 26–29 (1955)

[Šik90] Z. Šikić, Taylor's theorem. Int. J. Math. Educ. Sci. Technol. **21**(1), 111–115 (1990)

[SWa] N.J.A. Sloane, E.W. Weisstein, The regular paper-folding sequence (or dragon curve sequence). https://oeis.org/A014577

[Son94] J. Sondow, Analytic continuation of Riemann's zeta function and values at negative integers via Euler's transformation of series. Proc. Am. Math. Soc. **120**(2), 421–424 (1994)

[SWb] J. Sondow, E. Weisstein, Riemann zeta function. http://mathworld.wolfram.com/RiemannZetaFunction.html

Chapter 5
Oscillating Finite Sums

> *One cannot escape the feeling that these mathematical formulas*
> *have an independent existence and an intelligence of their own,*
> *that they are wiser than we are, wiser even than their*
> *discoverers.*
>
> Heinrich Hertz (1857–1894)

Abstract In this chapter, we use the theory of summability of divergent series, presented earlier in Chap. 4, to derive the analogs of the Euler-Maclaurin summation formula for oscillating sums. These formulas will, in turn, be used to perform many remarkable deeds with ease. For instance, they can be used to derive analytic expressions for summable divergent series, obtain asymptotic expressions of oscillating series, and even accelerate the convergence of series by several orders of magnitude. Moreover, we will prove the notable fact that, as far as the foundational rules of summability calculus are concerned, summable divergent series behave exactly as if they were convergent.

In this chapter, we use the $\mathfrak{T}$ definition of series, presented earlier in Chap. 4, to derive the analogs of the Euler-Maclaurin summation formula for oscillating sums. These formulas will, in turn, be used to perform many remarkable deeds with ease. For instance, they can be used to derive analytic expressions for summable divergent series, obtain asymptotic expressions of oscillating series, and even accelerate the convergence of series by several orders of magnitude. Moreover, we will prove the notable fact that, as far as the foundational rules of summability calculus are concerned, summable divergent series behave exactly as if they were convergent.

We will first address the important case of alternating sums, in which we derive a variant of the Boole summation formula [BCM09]. After that, we generalize results to a broader class of oscillating sums using the notion of *periodic sign sequences*, which brings to light the periodic analogs of Berndt and Schoenfeld [BS75] that were mentioned earlier in Chap. 1.

© Springer International Publishing AG 2018
I. M. Alabdulmohsin, *Summability Calculus*,
https://doi.org/10.1007/978-3-319-74648-7_5

5.1   Alternating Finite Sums

We will begin our treatment of alternating finite sums with the following lemma.

Lemma 5.1 *Suppose we have an alternating series of the form $\sum_{k=0}^{\infty}(-1)^k g(k)$, where $g(k)$ is analytic at an open disc around each point in the ray $[0, \infty)$. Also, suppose that the infinite sum is defined in $\mathfrak{T}$ by a value $V \in \mathbb{C}$. Then, V is equivalent to the following formal expression:*

$$V = \sum_{r=0}^{\infty} \frac{N_r}{r!} g^{(r)}(0), \quad \text{where } N_r = \sum_{k=0}^{\infty}(-1)^k k^r, \tag{5.1.1}$$

using the convention $0^0 \doteq 1$. Here, all series are interpreted using the generalized definition $\mathfrak{T}$.

Proof Since $g(k)$ is regular at the origin, we can rewrite V using the following formal expression:

$$V = \sum_{k=0}^{\infty}(-1)^k \sum_{r=0}^{\infty} \frac{g^{(r)}(0)}{r!} k^r \tag{5.1.2}$$

Rearranging:

$$V = \sum_{r=0}^{\infty} \frac{g^{(r)}(0)}{r!} \sum_{k=0}^{\infty}(-1)^k k^r \tag{5.1.3}$$

By definition of N_r, we have:

$$V = \sum_{r=0}^{\infty} \frac{N_r}{r!} g^{(r)}(0), \tag{5.1.4}$$

which is the statement of the lemma. $\square$

Lemma 5.1 is not very useful by itself. It simply provides us with an alternative method of computing the $\mathfrak{T}$ value of divergent series. However, it can be used to derive the analog of the Euler-Maclaurin summation formula for alternating sums as shown next.

Theorem 5.1 *Suppose we have a simple finite sum of the form $f(n) = \sum_{k=0}^{n-1}(-1)^k g(k)$, where $\sum_{k=0}^{\infty}(-1)^k g(k)$ is defined in $\mathfrak{T}$ by a value $V \in \mathbb{C}$. Then, its unique natural generalization $f_G(n)$ that is consistent with the successive polynomial approximation method of Theorem 2.1 is formally given by:*

$$f_G(n) = \sum_{r=0}^{\infty} \frac{N_r}{r!} \left(g^{(r)}(0) + (-1)^{n+1} g^{(r)}(n) \right) \tag{5.1.5}$$

In addition, we have for all $n \in \mathbb{C}$:

$$f_G(n) = \sum_{k=0}^{\infty}(-1)^k g(k) - \sum_{k=n}^{\infty}(-1)^k g(k) \qquad (5.1.6)$$

All infinite sums are interpreted using the generalized definition $\mathfrak{T}$.[1]

Proof We will first prove that Eq. 5.1.5 implies Eq. 5.1.6. By Lemma 5.1, we have Eq. 5.1.7, where both sums are interpreted using the generalized definition $\mathfrak{T}$.

$$\sum_{k=n}^{\infty}(-1)^k g(k) = (-1)^n \sum_{k=0}^{\infty}(-1)^k g(k+n) = (-1)^n \sum_{r=0}^{\infty} \frac{N_r}{r!} g^{(r)}(n) \qquad (5.1.7)$$

However:

$$\sum_{r=0}^{\infty} \frac{N_r}{r!}\left(g^{(r)}(0) + (-1)^{n+1} g^{(r)}(n)\right) = \sum_{k=0}^{\infty}(-1)^k g(k) - \sum_{k=n}^{\infty}(-1)^k g(k) \qquad (5.1.8)$$

Therefore, Eq. 5.1.6 indeed holds. This shows that summable divergent series behave as if they were convergent.

Second, we want to show that the generalized function $f_G(n)$ given by Eq. 5.1.5 is formally equivalent to the Euler-Maclaurin summation formula, which was derived in Chap. 2 using the successive polynomial approximation method of Theorem 2.1. To see this, we apply Eq. 5.1.5 on *non-alternating* simple finite sums to obtain:

$$\sum_{k=0}^{n-1} g(k) = \sum_{r=0}^{\infty} \frac{N_r}{r!}\left[\frac{d^r}{dx^r}\left(e^{i\pi x} g(x)\right)\Big|_{x=0} + (-1)^{n+1} \frac{d^r}{dx^r}\left(e^{i\pi x} g(x)\right)\Big|_{x=n}\right] \qquad (5.1.9)$$

Here, $i = \sqrt{-1}$. However, using symbolic methods, we know that:

$$\frac{d^r}{dx^r}\left(e^{i\pi x} g(x)\right) = e^{i\pi x}\left(i\pi + D_x\right)^r \qquad (5.1.10)$$

Here, D_x is the differential operator of the function $g(x)$, i.e. $D_x^r = g^{(r)}(x)$, and the expression $(i\pi + D_x)^r$ is to be interpreted by formally applying the binomial theorem on D_x. For example, $(i\pi + D_x)^2 = (i\pi)^2 D_x^0 + 2i\pi D_x^1 + D_x^2$. Therefore, Eq. 5.1.9 can be rewritten as:

$$\sum_{k=0}^{n-1} g(k) = \sum_{r=0}^{\infty} \frac{N_r}{r!}(i\pi + D_0)^r - \sum_{r=0}^{\infty} \frac{N_r}{r!}(i\pi + D_n)^r \qquad (5.1.11)$$

---

[1] The summation formula of Theorem 5.1 is similar to, but is different from, Boole's summation formula. Here, Boole's summation formula states that $\sum_{k=a}^{n}(-1)^k g(k) =$ $\frac{1}{2}\sum_{k=0}^{\infty} \frac{E_k(0)}{k!}\left((-1)^n f^{(k)}(n+1) + (-1)^a f^{(k)}(a)\right)$, where $E_k(x)$ are the Euler polynomials [BCM09, BBD89, Jor65].

Later in Eq. 5.1.14, we will show that $\frac{1}{1+e^x} = \sum_{r=0}^{\infty} \frac{N_r}{r!} x^r$. Using symbolic methods, we, therefore, have:

$$\sum_{k=0}^{n-1} g(k) = \frac{1}{1 + e^{i\pi + D_0}} - \frac{1}{1 + e^{i\pi + D_n}} = \frac{1}{1 - e^{D_0}} + \frac{1}{1 - e^{D_n}} \qquad (5.1.12)$$

Using the series expansion of the function $\frac{1}{1-e^x}$, we deduce that Eq. 5.1.13 holds, where B_r are Bernoulli numbers.

$$\sum_{k=0}^{n-1} g(k) = \sum_{r=0}^{\infty} \frac{B_r}{r!} (D_n^{r-1} - D_a^{r-1}) \qquad (5.1.13)$$

However, this last equation is precisely the Euler-Maclaurin summation formula. Thus, the generalized function $f_G(n)$ to alternating simple finite sums given by Eqs. 5.1.5 and 5.1.6 is indeed the unique most natural generalization of alternating sums that is consistent with the successive polynomial approximation method of Theorem 2.1, which completes the proof of the theorem.   □

As shown earlier, the constants N_r are given by the closed-form expression in Eq. 4.1.10. For example, the first ten terms are $N_r = \{\frac{1}{2}, -\frac{1}{4}, 0, \frac{1}{8}, 0, -\frac{1}{4}, 0, \frac{17}{16}, 0, -\frac{31}{4}, \ldots\}$. A generating function for the constants N_r can be deduced from Lemma 5.1 if we let $g(k) = e^{kx}$, which yields:

$$\sum_{k=0}^{\infty} (-1)^k e^{kx} = \frac{1}{1 + e^x} = \sum_{r=0}^{\infty} \frac{N_r}{r!} x^r \qquad (5.1.14)$$

The constants N_r can be thought of as the alternating analog of Bernoulli numbers. However, it can be shown that one defining property of Bernoulli numbers is the fact that they are the unique solutions to the functional equation:

$$g'(n-1) = \sum_{r=0}^{\infty} \frac{B_r}{r!} [g^{(r)}(n) - g^{(r)}(n-1)] \qquad (5.1.15)$$

From Theorem 5.1, we can immediately deduce a similar defining property of the constants N_r. In particular, if we consider the function $\sum_{k=0}^{n}(-1)^k g(k) - \sum_{k=0}^{n-1}(-1)^k g(k)$, we have that N_r are the solutions to the following functional equation:

$$g(n-1) = \sum_{r=0}^{\infty} \frac{N_r}{r!} [g^{(r)}(n) + g^{(r)}(n-1)] \qquad (5.1.16)$$

By formally applying the Taylor series expansion of both sides of Eq. 5.1.16, we arrive at the following recursive method for computing N_r:

$$N_r = -\frac{1}{2} \sum_{k=0}^{r-1} \binom{r}{k} N_k \qquad \text{if } r > 0, \text{ and } N_0 = \frac{1}{2} \qquad (5.1.17)$$

One immediate implication of Theorem 5.1 is that it gives us closed-form formulas for *alternating* power sums. For example, we have:

$$\sum_{k=0}^{n-1}(-1)^k = \frac{1}{2} - \frac{(-1)^n}{2} \qquad (5.1.18)$$

$$\sum_{k=0}^{n-1}(-1)^k k = -\frac{1}{4} + (-1)^{n+1}\frac{2n+1}{4} \qquad (5.1.19)$$

Note that in both examples, the $\mathfrak{T}$ sequence limit of $(-1)^n$ and $(-1)^n n$ as $n \to \infty$ is zero by Proposition 4.3 so the value of the divergent series are consistent with the $\mathfrak{T}$ sequence limit in these equations. In general, we have for all integers $r \geq 0$ the following analog to the Bernoulli-Faulhaber formula:

$$\sum_{k=0}^{n-1}(-1)^k k^r = N_r + (-1)^{n+1}\sum_{k=0}^{r}\binom{r}{k}N_k\,n^{r-k} \qquad (5.1.20)$$

Corollary 5.1 *Suppose $g(n)$ has a finite polynomial order m, i.e. $g^{(m+1)}(n) \to 0$ as $n \to \infty$, and suppose that the value of $\sum_{k=0}^{\infty}(-1)^k g(k)$ has a value V under the definition $\mathfrak{T}$. Then, V can be evaluated using the following limit:*

$$V = \lim_{n\to\infty}\Big\{\sum_{k=0}^{n-1}(-1)^k g(k) + (-1)^n\sum_{r=0}^{m}\frac{N_r}{r!}g^{(r)}(n)\Big\} \qquad (5.1.21)$$

Taking higher order terms of the expression $\sum_{r=0}^{m}\frac{N_r}{r!}g^{(r)}(n)$ will improve the speed of convergence. Alternatively, and under the stated conditions, the simple finite sum $\sum_{k=0}^{n-1}(-1)^k g(k)$ is asymptotically given by the following expression, where the error term vanishes as n goes to infinity.

$$\sum_{k=0}^{n-1}(-1)^k g(k) \sim V + (-1)^{n+1}\sum_{r=0}^{m}\frac{N_r}{r!}g^{(r)}(n) \qquad (5.1.22)$$

Proof Follows immediately from Lemma 5.1 and Theorem 5.1. ☐

Corollary 5.1 provides us with a simple method of obtaining asymptotic expressions to alternating series, assigning natural values to divergent alternating series, and accelerating the convergence of alternating series as well. It can even allow us to *derive* analytic expressions of divergent alternating series in some cases.

For example, suppose we would like to apply the generalized definition $\mathfrak{T}$ to the alternating series $\sum_{k=0}^{\infty}(-1)^k \log(1+k)$. Numerically, if we use the summability method χ with $n = 100$ or $n = 1000$, we get -0.2264 and -0.2259 respectively. Using larger values of n show that the divergent series is summable to around -0.2258 (up to four decimal places). Using Corollary 5.1, we can derive the exact value of such a divergent series as follows. First, we note that $\log n$ is asymptotically of a finite differentiation order zero. Thus, we have:

$$\sum_{k=0}^{\infty}(-1)^k \log(1+k) = \lim_{n\to\infty}\left\{\sum_{k=1}^{2n}(-1)^{k+1}\log k + \frac{\log(2n)}{2}\right\}$$

$$= \lim_{n\to\infty}\left\{\sum_{k=1}^{2n}\log k - 2\sum_{k=1}^{n}\log(2k) + \frac{\log(2n)}{2}\right\}$$

$$= \lim_{n\to\infty}\left\{\log(2n)! - 2n\log 2 - 2\log n! + \frac{\log(2n)}{2}\right\}$$

$$= \frac{1}{2}\log\frac{2}{\pi}$$

Here, we have used the Stirling approximation in the last step [Rob55]. Indeed, we have $\frac{1}{2}\log\frac{2}{\pi} = -0.2258$, which agrees numerically with the summability method χ.

Similarly, if we apply Corollary 5.1 to the divergent series $\sum_{k=0}^{\infty}(-1)^k \log(k!)$, whose polynomial order is $m = 1$, we obtain Eq. 5.1.23, which can be confirmed numerically quite readily using the summability method χ.

$$\sum_{k=0}^{\infty}(-1)^k \log(k!) = \frac{1}{4}\log\frac{\pi}{2} \qquad (5.1.23)$$

As stated earlier, we can also employ Corollary 5.1 to accelerate the "convergence" of alternating series, including alternating *divergent* series for which the generalized definition $\mathfrak{T}$ is implied. For example, suppose we would like to compute $\sum_{k=0}^{\infty}(-1)^k \log(1+k)$ using the summability method χ. If we use $n = 100$, we obtain a figure that is accurate up to three decimal places only, which is expected due to the fact that the summability method converges only algebraically. If, on the other hand, we compute the divergent series using Corollary 5.1, where $m = 1$ and $n = 100$, we obtain a figure that is accurate up to seven decimal places! Here, we chose $m = 1$ instead of $m = 0$ to accelerate the convergence.

Moreover, we can obviously use Corollary 5.1 to *accelerate* the convergence of alternating series. For example, applying Corollary 5.1 to the convergent series $\sum_{k=0}^{\infty}\frac{(-1)^k}{(1+k)^2}$ using $n = 100$ and $m = 1$ yields a value of 0.8224670334, which is accurate up to ten decimal places! This is quite remarkable given the fact that we have only used 100 terms in such a slowly converging sum. Of course, choosing higher values of m would accelerate the convergence even more.[2]

One example where Corollary 5.1 can be employed to obtain asymptotic expressions is the *second factorials* function given by $f(n) = \prod_{k=1}^{n} k^{k^2}$. Similar to

---

[2]It is worth mentioning that many algorithms exist for accelerating the convergence of alternating series, some of which can sometimes yield several digits of accuracy per iteration. Interestingly, some of those algorithms, such as the one proposed in [CVZ00], were also found to be capable of "correctly" summing some divergent alternating series as well.

earlier approaches, we will first find an asymptotic expression to $\log f(n)$. However, this is easily done using the Euler-Maclaurin summation formula but its leading constant is unfortunately unknown in closed-form. To find an exact expression of that constant, we use the fact that $\sum_{k=1}^{\infty}(-1)^k k^2 \log k = 7\zeta(3)/(4\pi^2)$, which can be deduced immediately from the well-known analytic continuation of the Riemann zeta function. Here, we know by Eq. 4.1.4 that the aforementioned assignment indeed holds when interpreted using the generalized definition $\mathfrak{T}$. Now, we employ Corollary 5.1, which yields the asymptotic expression of the second factorials function given in Eq. 5.1.24, with a ratio that goes to unity as $n \to \infty$. The asymptotic expression in Eq. 5.1.24 is mentioned in [CS00].

$$\prod_{k=1}^{n} k^{k^2} \sim e^{\frac{\zeta(3)}{4\pi^2}} n^{\frac{2n^3+3n^2+n}{6}} e^{\frac{n}{12}-\frac{n^3}{9}} \tag{5.1.24}$$

The formal expressions for the $\mathfrak{T}$ value of divergent series in Lemma 5.1 can be used to derive the Euler summation method.

Proposition 5.1 (The Euler Sum Revisited) *Let $V_1 = \sum_{k=0}^{\infty}(-1)^k g(k)$ when interpreted by $\mathfrak{T}$. Also, let $V_2 = \sum_{k=0}^{\infty}(-1)^k \sum_{j=0}^{k-1} g(j)$ when interpreted by $\mathfrak{T}$. If both V_1 and V_2 exist, then $V_1 = -2V_2$.*

Proof We have by Theorem 5.1 that V_1 is given by:

$$V_1 = \sum_{k=0}^{n-1}(-1)^k g(k) + (-1)^n \sum_{r=0}^{\infty} \frac{N_r}{r!} g^{(r)}(n) \tag{5.1.25}$$

Similarly, we have:

$$V_2 = \sum_{k=0}^{n-1}(-1)^k G(k) + (-1)^n \sum_{r=0}^{\infty} \frac{N_r}{r!} G^{(r)}(n), \quad G(k) = \sum_{j=0}^{k-1} g(j) \tag{5.1.26}$$

However, we know by the basic rules of summability calculus in Chap. 2 that the following holds for all $r \geq 0$:

$$G^{(r)}(n) = \sum_{k=0}^{n-1} g^{(r)}(k) + G^{(r)}(0) \tag{5.1.27}$$

Plugging Eq. 5.1.27 into Eq. 5.1.26 yields:

$$V_2 = \sum_{k=0}^{n-1}(-1)^k G(k) + (-1)^n \sum_{r=0}^{\infty} \frac{N_r}{r!} G^{(r)}(0) + (-1)^n \sum_{r=0}^{\infty} \frac{N_r}{r!} \sum_{j=0}^{n-1} g^{(r)}(j) \tag{5.1.28}$$

We know by Lemma 5.1 that:

$$\sum_{r=0}^{\infty} \frac{N_r}{r!} G^{(r)}(0) = V_2 \tag{5.1.29}$$

Therefore, Eq. 5.1.28 can be rewritten as:

$$V_2 = \sum_{k=0}^{n-1}(-1)^k\, G(k) + (-1)^n V_2 + (-1)^n \sum_{r=0}^{\infty} \frac{N_r}{r!} \sum_{j=0}^{n-1} g^{(r)}(j) \qquad (5.1.30)$$

The expression holds for all n. We set $n = 1$ to obtain:

$$2V_2 = -\sum_{r=0}^{\infty} \frac{N_r}{r!} g^{(r)}(0) = -V_1 \qquad (5.1.31)$$

Therefore, we have the desired result.   □

One example of Proposition 5.1 was already demonstrated for the infinite sum $\sum_{k=0}^{\infty}(-1)^k \log(k!)$, where we showed in Eq. 5.1.23 that the $\mathfrak{T}$ value of such a divergent series is $\log(\pi/2)/4$. However, comparing this with the $\mathfrak{T}$ value of the divergent series $\sum_{k=1}^{\infty}(-1)^{k+1}\log k$ given earlier, we see that the Lemma holds as expected.

In addition, since $\sum_{k=1}^{\infty}(-1)^k 1/(1+k) = \log 2$, we deduce from Proposition 5.1 that:

$$\sum_{k=0}^{\infty}(-1)^k H_{1+k} = \frac{\log 2}{2} \qquad (5.1.32)$$

Therefore, we have by linearity of $\mathfrak{T}$:

$$\sum_{k=0}^{\infty}(-1)^k (H_{1+k} - \log(1+k) - \gamma) = \frac{\log \pi - \gamma}{2} \qquad (5.1.33)$$

Here, we have used the values of divergent series to compute a convergent sum!

Now, Theorem 5.1 can be readily generalized to composite alternating sums as the following theorem shows.

Theorem 5.2 *Given a composite finite sum of the form* $f(n) = \sum_{k=0}^{n-1}(-1)^k g(k,n)$, *then its unique natural generalization* $f_G(n)$ *that is consistent with the successive polynomial approximation method of Theorem 2.1 is formally given by:*

$$f_G(n) = \sum_{r=0}^{\infty} \frac{N_r}{r!} \left[\frac{\partial^r}{\partial t^r} g(t,n) \Big|_{t=0} + (-1)^{n+1} \frac{\partial^r}{\partial t^r} g(t,n) \Big|_{t=n} \right] \qquad (5.1.34)$$

Proof Similar to the proof of Lemma 3.1 and Theorem 3.1.   □

Theorem 5.2 gives a simple method for deriving closed-form expressions and/or asymptotic expansions of alternating composite finite sums. For example, suppose $f(n)$ is given by the composite alternating sum $f(n) = \sum_{k=0}^{n-1}(-1)^k \log(1 + \frac{k}{n})$, and suppose that we wish to find its asymptotic value as n tends to infinity. Using Theorem 5.2, we have:

$$\sum_{k=0}^{n-1}(-1)^k \log\left(1+\frac{k}{n}\right) \sim \frac{N_0}{0!}\left[\log\left(1+\frac{t}{n}\right)\Big|_{t=0} + (-1)^{1+n}\log\left(1+\frac{t}{n}\right)\Big|_{t=n}\right]$$

$$= (-1)^{n+1}\frac{\log 2}{2},$$

which can be verified numerically.

Exercise 5.1 Derive the summation formula for alternating sums given in Theorem 5.1 from the Euler-Maclaurin summation formula *directly* upon using the identity $\sum_{k=0}^{2n-1}(-1)^k g(k) = \sum_{k=0}^{2n-1} g(k) - 2\sum_{k=0}^{n-1} g(2k+1)$. Whereas the Boole summation formula can be derived from the Euler-Maclaurin formula, our approach using analytic summability theory yields a more general framework that can be easily extended to oscillating sums as will be shown later in this chapter.

Exercise 5.2 Show that the constants N_r are given by the expression $N_r = (1 - 2^{1+r})B_{r+1}/(r+1)$, where B_r are the Bernoulli numbers. (Hint: use Eq. 4.1.11 in Chap. 4 and the fact that $\zeta(-s) = -B_{s+1}/(s+1)$ for $s \geq 1$.)

Exercise 5.3 Use Proposition 5.1 to prove that $\sum_{k=1}^{\infty}(-1)^{k+1}\log S(k) = \frac{\log(2/\pi)}{8} \approx -0.0564$, where $S(k) = k!\,S(k-1)$ is the superfactorial function. Verify this result numerically using the summability method χ.

Exercise 5.4 Prove that the alternating simple finite sum $\sum_{k=0}^{n-1}(-1)^{-k}g(k)$ is formally given by:

$$\sum_{k=0}^{n-1}(-1)^{-k}g(k) = \sum_{r=0}^{\infty}\frac{N_r}{r!}\left(g^{(r)}(0) - (-1)^{-n}g^{(r)}(n)\right) \qquad (5.1.35)$$

Note that the sign sequence is $(-1)^{-k}$, not $(-1)^k$. Conclude that $\sum_{k=0}^{n-1}(-1)^{-k}g(k) \neq \sum_{k=0}^{n-1}(-1)^k g(k)$, in general, when $n \in \mathbb{C}$. Nevertheless, they are equal to each other if $n \in \mathbb{N}$.

Exercise 5.5 Use Theorem 5.2 to prove that:

$$\sum_{k=0}^{n}(-1)^k(k+n) = \frac{n}{2} + (-1)^n n + \frac{(-1)^n - 1}{4} \qquad (5.1.36)$$

Show that the same expression can be proved differently by splitting the sum into two alternating sums $\sum_{k=0}^{n}(-1)^k(k+n) = \sum_{k=0}^{n}(-1)^k k + n\sum_{k=0}^{n}(-1)^k$, and by using the results in Eqs. 5.1.18 and 5.1.19.

Exercise 5.6 Prove that the Bernoulli numbers are the unique solutions to the function equation:

$$g'(n-1) = \sum_{r=0}^{\infty} \frac{B_r}{r!} \left[g^{(r)}(n) - g^{(r)}(n-1) \right] \tag{5.1.37}$$

Also, prove that the constants N_r of Theorem 5.2 are the unique solutions to the functional equation:

$$g(n-1) = \sum_{r=0}^{\infty} \frac{N_r}{r!} \left[g^{(r)}(n) - g^{(r)}(n-1) \right] \tag{5.1.38}$$

5.2 Oscillating Sums: The General Case

The approach employed in the previous section for alternating sums can be readily extended to finite sums of the form $\sum_{k=0}^{n-1} e^{i\theta k} g(k)$, which is stated in the following theorem.

5.2.1 Complex Exponentials

Theorem 5.3 *If a simple finite sum is of the form* $f(n) = \sum_{k=0}^{n-1} e^{i\theta k} g(k)$, *where* $0 < \theta < 2\pi$, *then its unique natural generalization* $f_G(n)$ *that is consistent with the successive polynomial approximation method of Theorem 2.1 is formally given by:*

$$f_G(n) = \sum_{r=0}^{\infty} \frac{\Theta_r}{r!} \left[g^{(r)}(0) - e^{i\theta n} g^{(r)}(n) \right], \quad \text{where } \Theta_r = \sum_{k=0}^{\infty} e^{i\theta k} k^r$$

Here, infinite sums are interpreted using the generalized definition $\mathfrak{T}$. *In addition, we have:*

$$f_G(n) = \sum_{k=0}^{\infty} e^{i\theta k} g(k) - \sum_{k=n}^{\infty} e^{i\theta k} g(k) \tag{5.2.1}$$

Also, if g is of a finite polynomial order m, then the series $\sum_{k=0}^{\infty} e^{i\theta k} g(k)$ *is equal to the limit:*

$$\sum_{k=0}^{\infty} e^{i\theta k} g(k) = \lim_{n\to\infty} \left\{ \sum_{k=0}^{n-1} e^{i\theta k} g(k) + \sum_{r=0}^{m} \frac{\Theta_r}{r!} e^{i\theta n} g^{(r)}(n) \right\}$$

Proof Similar to the proofs of Theorem 5.1 and Corollary 5.1.   □

If $\theta = \pi$, we have $\Theta_r = N_r$. To find a generating function for Θ_r in general, we set $g(k) = e^{kx}$, which yields the generating function:

$$\sum_{k=0}^{\infty} e^{i\theta k} e^{kx} = \frac{1}{1 - e^{x+i\theta}} = \sum_{r=0}^{\infty} \frac{\Theta_r}{r!} x^r \qquad (5.2.2)$$

Theorem 5.3 paves the road for the general case of oscillating finite sums. It can be extended to composite oscillatory sums as well in a manner that is similar to Theorem 5.2, but we leave this extension as an exercise to the reader.

More than two centuries ago, and using arguments that were similar to Leibniz's, Daniel Bernoulli argued that periodic oscillating sums should be interpreted probabilistically. For instance, he stated that the Grandi series $\sum_{k=0}^{\infty}(-1)^k$ should be assigned the value of $\frac{1}{2}$ because the partial sums were equal to one if n was even and were equal to zero otherwise; therefore, the *expected value* of the infinite sum was $\frac{1}{2}$. Following the same reasoning, Bernoulli suggested that the divergent sum $1+0-1+1+0-1\ldots$ should be assigned the value of $\frac{2}{3}$ [Har49, San06]. Remarkably, such values are indeed consistent with many natural summability methods, such as the Abel summability method and the summability method χ.

Almost a century later, Cesàro expanded on the same principle by proposing his well-known summability method in 1890 that is given by Eq. 5.2.3 [Har49]. The intuition behind Cesàro summability method, however, is not Bernoulli's probabilistic interpretation but, rather, the notion of what is currently referred to as the *Cesàro mean*. Simply stated, if an infinite sum $\sum_{k=0}^{\infty} g(k)$ converges to some value V, then the partial sums $\sum_{k=0}^{j} g(k)$ approach V for sufficiently large values of j. Thus, $\sum_{j=0}^{n} \sum_{k=0}^{j} g(k)$ will approach $n\, V$ as $n \to \infty$, which makes Eq. 5.2.3 a very "natural" definition for divergent series, in general, whenever the limit exists. It is straightforward to observe that the Cesàro mean is consistent with, but is more powerful than, Bernoulli's probabilistic interpretation.

$$\sum_{k=0}^{\infty} g(k) \triangleq \lim_{n\to\infty} \left\{ \frac{1}{n} \sum_{j=1}^{n} \sum_{k=0}^{j} g(k) \right\} \qquad (5.2.3)$$

In this section, we prove that the intuitive reasoning of Bernoulli is indeed correct when series are interpreted using the generalized definition $\mathfrak{T}$. This was established more than once in Chap. 4; it was proved directly in Exercise 4.1 and indirectly via the consistency of $\mathfrak{T}$ with the Cesàro summability method. Nonetheless, the results presented in this section will allow us to deduce an alternative direct proof to Bernoulli's probabilistic interpretation. We use this proof technique to derive a recurrence relation for the constants Θ_r of Theorem 5.3.

Throughout the sequel, we will restrict our attention to oscillating sums of the form $\sum_{k=0}^{n-1} s_k\, g(k)$, where $g(k)$ is regular at the origin and s_k is an arbitrary *periodic* sequence of numbers that will be referred to as *the sign sequence*. Common examples of sign sequences include $(1, 1, 1, \ldots)$ in ordinary summation, $(1, -1, 1, -1, \ldots)$ in the case of alternating sums, and $(e^{i2\pi/n}, e^{i4\pi/n}, e^{i6\pi/n}, \ldots)$ in harmonic analysis, and so on. It is important to keep in mind that the sign sequence is assumed to be periodic, but it does not have to be analytic.

5.2.2 Arbitrary Sign Sequences

Looking into the proofs of Lemma 5.1 and Theorem 5.1, we note that if a similar approach is to be employed for oscillating sums of the form $\sum_{k=0}^{n-1} s_k \, g(k)$, then the series $\sum_{k=0}^{\infty} s_k \, k^r$ has to be well-defined in $\mathfrak{T}$ for all $r \geq 0$. Thus, our first point of departure is to ask for which periodic sign sequences s_k do the series $\sum_{k=0}^{\infty} s_k \, k^r$ exist in $\mathfrak{T}$ for all integers $r \geq 0$. Proposition 5.2 provides us with the complete answer.

Proposition 5.2 *Given a periodic sign sequence s_k with period p, i.e. $s_{k+p} = s_k$, then the series $\sum_{k=0}^{\infty} s_k \, k^r$ are well-defined in $\mathfrak{T}$ for all integers $r \geq 0$ if and only if $\sum_{k=0}^{p-1} s_k = 0$. That is, the average of the values of the sign sequence s_k for any cycle of length p is equal to zero.*

Proof To prove that the condition $\sum_{k=0}^{p-1} s_k = 0$ is necessary, we set $r = 0$, which yields by stability:

$$\sum_{k=0}^{\infty} s_k = \sum_{k=0}^{p-1} s_k + \sum_{k=p}^{\infty} s_k = \sum_{k=0}^{p-1} s_k + \sum_{k=0}^{\infty} s_k \qquad (5.2.4)$$

Therefore, if $\sum_{k=0}^{\infty} s_k$ is summable to some value $V \in \mathbb{C}$, then $\sum_{k=0}^{p-1} s_k$ must be equal to zero.

To show that the condition is sufficient, we first note by Taylor's theorem that:

$$\frac{x-1}{x^p - 1} = 1 - x + x^p - x^{p+1} + x^{2p} - x^{2p-1} \cdots \qquad (5.2.5)$$

Therefore, if we define a function $f_S(x)$ by:

$$f_S(x) = \frac{x-1}{x^p - 1}\left(s_0 + (s_0 + s_1)\, x + \cdots + (s_0 + s_1 + \ldots + s_{p-2})\, x^{p-2}\right), \qquad (5.2.6)$$

where $s_{p-1} = -\sum_{k=0}^{p-2} s_k$, the Taylor series expansion of $f_S(x)$ is given by:

$$f_S(x) = \sum_{k=0}^{\infty} s_k \, x^k \qquad (5.2.7)$$

Because $f_S(x)$ is analytic in an open disc around each point in the domain $[0, 1]$, then the $\mathfrak{T}$ definition of $\sum_{k=0}^{\infty} s_k \, k^r$ yields some linear combination of higher order derivatives $f_S^{(m)}(1)$ (see Proposition 5.3). Therefore, they are also well-defined in $\mathfrak{T}$. $\square$

Proposition 5.3 *Given a periodic sign sequence s_k with period p, where $\sum_{k=0}^{p-1} s_k = 0$, let S_r be given by $S_r = \sum_{k=0}^{\infty} s_k k^r$, where the divergent series is interpreted using $\mathfrak{T}$. Then, S_r can be exactly computed using the following recurrence identity:*

$$\sum_{k=0}^{r-1} \binom{r}{k} p^{r-k} S_k + \sum_{0}^{p-1} s_k k^r = 0, \qquad S_0 = \frac{1}{p} \sum_{k=0}^{p-1} \sum_{i=0}^{k} s_i \qquad (5.2.8)$$

Proof Again, the proof rests on the fact that the $\mathfrak{T}$ definition of series $\sum_{k=0}^{\infty} s_k k^r$ is stable, which implies:

$$\sum_{k=0}^{\infty} s_k k^r = \sum_{k=0}^{p-1} s_k k^r + \sum_{k=0}^{\infty} s_k (k+p)^r \qquad (5.2.9)$$

Expanding the factors $(k+p)^r$ using the Binomial Theorem and rearranging the terms yields the desired result. □

Now, knowing that any periodic sequence s_k with period p that satisfies the condition of Proposition 5.2 imply that $\sum_{k=0}^{\infty} s_k k^r$ exists in $\mathfrak{T}$ for all integers $r \geq 0$, and given that we can compute the exact values of $\sum_{k=0}^{\infty} s_k k^r$ for all $r \geq 0$ using Proposition 5.3, we are ready to generalize all of the results of the previous section.

First, given an arbitrary periodic sign sequence s_k, we compute its average $\tau = \sum_{k=0}^{p-1} s_k$ and split the sum into:

$$\sum_{k=0}^{n-1} s_k g(k) = \tau \sum_{k=0}^{n-1} g(k) + \sum_{k=0}^{n-1} (s_k - \tau) g(k)$$

The first term is a direct simple finite sum that can be analyzed using the earlier results of summability calculus in Chap. 2. The second term, on the other hand, satisfies the condition $\sum_{k=0}^{p-1} (s_k - \tau) = 0$ so Propositions 5.2 and 5.3 both hold.

Theorem 5.4 *Given a simple finite sum of the form $f(n) = \sum_{k=0}^{n-1} s_k g(k)$, where $s_k = (s_0, s_1, \ldots)$ is a periodic sign sequence with period p that satisfies the condition $\sum_{k=0}^{p-1} s_k = 0$. Then, for all $n \in \mathbb{N}$, $f(n)$ is formally given by:*

$$f(n) = \sum_{r=0}^{\infty} \frac{1}{r!} \left[S_r(0) g^{(r)}(0) - S_r(n \bmod p) g^{(r)}(n) \right] \qquad (5.2.10)$$

Here, $S_r(x) = \sum_{k=0}^{\infty} s_{k+x} k^r$ can be computed exactly using Proposition 5.3. In addition, if $g(n)$ is of a finite polynomial order m, then:

$$\sum_{k=0}^{\infty} s_k g(k) = \lim_{n \to \infty} \left\{ \sum_{k=0}^{pn-1} s_k g(k) + \sum_{r=0}^{m} \frac{S_r(0)}{r!} g^{(r)}(pn) \right\} \qquad (5.2.11)$$

Proof Similar to the proofs of Theorem 5.1 and Corollary 5.1.   □

Remark 5.1 Note that we cannot consider $f(n)$ given by Eq. 5.2.10 to be the natural generalization $f_G(n)$ since it is, so far, only defined for $n \in \mathbb{N}$ due to the use of the modulus after division operation. This limitation will be resolved shortly using the Discrete Fourier Transform (DFT).

Again, Theorem 5.4 allows us to derive the asymptotic expansions of oscillating finite sums and to accelerate the convergence of series. For instance, suppose $s_k = (0, 1, 0, -1, 0, 1, 0, \ldots)$, i.e. $p = 4$ and $s_0 = 0$, $s_1 = 1$, $s_2 = 0$, $s_3 = -1$, ... Then:

$$S_0(0) = 0 + 1 + 0 - 1 + 0 + 1 + 0 - 1 + \cdots = \frac{0 + 1 + 1 + 0}{4} = \frac{1}{2} \quad (5.2.12)$$

Similarly, $S_1(0)$ can be computed using Proposition 5.3, which yields $S_1(1) = -\frac{1}{2}$. Therefore, if we wish to obtain a method of evaluating the convergent sum $\sum_{k=0}^{\infty} \frac{s_k}{1+k} = \frac{1}{1} - \frac{1}{3} + \frac{1}{5} - \ldots = \frac{\pi}{4}$ with a cubic-convergence speed, we choose $m = 1$ in Eq 5.2.11. For instance, choosing $n = 100$ yields:

$$\sum_{k=0}^{\infty} \frac{s_k}{1+k} \approx \sum_{k=0}^{99} \frac{s_k}{1+k} + \frac{S_0(0)}{101} - \frac{S_1(0)}{101^2} = 0.78539817 \quad (5.2.13)$$

Here, we obtain a figure that is accurate up to seven decimal places. This is indeed quite remarkable given that we have only used 100 terms in the series. Otherwise, if we compute the sum directly, we would need to, approximately, evaluate 500,000 terms to achieve the same level of accuracy!

5.2.3   The Discrete Fourier Transform

Finally, to obtain the unique natural generalization $f_G(n)$ for simple finite sums of the form $f(n) = \sum_{k=0}^{n-1} s_k g(k)$ in which s_k is a periodic sign sequence, we use the Discrete Fourier Transform (DFT) as the following lemma shows.

Lemma 5.2 *Suppose we have a periodic sign sequence* $s_k = (s_0, s_1, \ldots)$ *with period p. Then, s_k can be generalized to all complex values of k using the Discrete Fourier Transform (DFT). More precisely, we have:*

$$s_k = \sum_{m=0}^{p-1} v_m e^{i\frac{mk}{p}}, \quad \text{where} \quad v_m = \frac{1}{p} \sum_{k=0}^{p-1} s_k e^{-i\frac{mk}{p}} \quad (5.2.14)$$

Proof By direct application of the Discrete Fourier Transform (DFT).   □

Remark 5.2 Intuitively, v_m is the *projection* of the sequence s_k on the sequence $\{e^{i\frac{mk}{p}}\}_k$ when both are interpreted as infinite-dimensional vectors and $\{e^{i\frac{mk}{p}}\}_k$ are orthogonal for different values of m. In addition, since s_k is periodic, the basis are complete. Note that v_0 is the average value of the of sign sequence s_k, i.e. the DC component in engineering terminology.

Corollary 5.2 *Given a simple finite sum of the form $f(n) = \sum_{k=0}^{n-1} s_k \, g(k)$, where $s_k = (s_0, s_1, \ldots)$ is a periodic sign sequence with period p, we have:*

$$f(n) = v_0 \sum_{k=0}^{n-1} g(k) + v_1 \sum_{k=0}^{n-1} e^{i\frac{k}{p}} g(k) + \cdots + v_{p-1} \sum_{k=0}^{n-1} e^{i\frac{(p-1)k}{p}} g(k) \qquad (5.2.15)$$

Here, v_m are defined in Lemma 5.2. In addition, the unique natural generalization $f_G(n)$ that is consistent with the successive polynomial approximation method of Theorem 2.1 is given by the sum of unique natural generalizations to all terms in the previous equation (see Theorem 5.3).

Proof By Lemma 5.2. □

Exercise 5.7 Use Proposition 5.2 and L'Hôpital's rule to show that if s_k is periodic with period p and $\sum_{k=0}^{p-1} s_k = 0$, then the $\mathfrak{T}$ value of $\sum_{k=0}^{\infty} s_k$ is equal to $\frac{1}{p} \sum_{k=0}^{p-1} \sum_{i=0}^{k} s_i$.

Exercise 5.8 Consider the sign sequence $s_k = (+1, +1, -1, -1, +1, +1, -1, -1, \ldots)$, where $p = 4$. Use the recursive identity in Proposition 5.3 to prove that $S_0(0) = 1$ and $S_1(0) = -\frac{1}{2}$. Apply the series acceleration method in Theorem 5.4 to evaluate the series $\sum_{k=0}^{\infty} s_k/(k+1)$. Knowing that $\sum_{k=0}^{\infty} s_k/(k+1) = \frac{\pi}{4} + \frac{\log 2}{2}$, show that:

1. By summing the first 4000 terms in the series, we obtain a figure that is accurate up to three decimal places only.
2. By adding the first term of the asymptotic expansion $S_r(0)/(4n+1)$, we produce a figure that is accurate up to seven decimal places.
3. By, further, adding the second term $-S_r(1)/(4n+1)^2$ of the asymptotic expansion, we improve the accuracy to 11 decimal places.

Therefore, series acceleration can be improved considerably using Theorem 5.4.

5.3 Infinitesimal Calculus on Oscillating Sums

One particular case in which analyzing simple finite sums of the form $\sum_{k=0}^{n-1} g(k)$ is relatively simple is when the series $\sum_{k=0}^{\infty} g(k)$ converges. In this section, we will show that if the series $\sum_{k=0}^{\infty} g(k)$ is well-defined under $\mathfrak{T}$, then the function $f(n)$ behaves with respect to infinitesimal calculus exactly as if it were a convergent function as $n \to \infty$, regardless of whether or not the limit $\lim_{n \to \infty} f(n)$ exists in the classical sense of the word.

In his notebooks, Ramanujan captured the basic idea that ties summability theory to infinitesimal calculus. In his words, every series has a constant c, which acts "like the center of gravity of the body." However, his definition of the constant c was imprecise and frequently led to incorrect conclusions [Ber85]. Assuming that such a constant exists, Ramanujan reasoned that fractional sums could be defined by:

$$\sum_{k=0}^{n-1} g(k) = \sum_{k=0}^{\infty} g(k) - \sum_{k=n}^{\infty} g(k) = c_g(0) - c_g(n) \tag{5.3.1}$$

Here, $c_g(x)$ is the "constant" of the series $\sum_{k=x}^{\infty} g(k)$. Next, Ramanujan deduced that:

$$\frac{d}{dn} \sum_{k=0}^{n-1} g(k) = \frac{d}{dn} \left(\sum_{k=0}^{\infty} g(k) - \sum_{k=n}^{\infty} g(k) \right)$$

$$= -\frac{d}{dn} c_g(n) = -\sum_{k=n}^{\infty} g'(k)$$

$$= \sum_{k=0}^{n-1} g'(k) - \sum_{k=0}^{\infty} g'(k) = \sum_{k=0}^{n-1} g'(k) - c_{g'}(0)$$

Therefore, we recover the differentiation rule of simple finite sums that we started with in Chap. 2.

This reasoning would have been correct if a precise and a consistent definition of the constant c existed. However, a consistent definition of c for *all* series cannot be attained. For instance, suppose we have the function $f(n) = \sum_{k=0}^{n-1}(g(k) + \tau)$ for some constant τ. Then, the last line in the earlier reasoning implies that $f'(n)$ is independent of τ, which is clearly incorrect because $f'(n) = \tau + \frac{d}{dn} \sum_{k=0}^{n-1} g(k)$.

In this section, we will show that Ramanujan's idea is correct but *only* when the series $\sum_{k=0}^{\infty} g(k)$ is well-defined under $\mathfrak{T}$. In this case, introducing a constant τ as mentioned above would violate the condition of summability so the method is well-defined.

We will begin our discussion with simple finite sums of the form $\sum_{k=0}^{n-1} e^{i\theta k} g(k)$ and state the general results afterwards.

Lemma 5.3 *Suppose we have a simple finite sum of the form $f(n) = \sum_{k=0}^{n-1} e^{i\theta k} g(k)$, where the infinite sum $\sum_{k=0}^{\infty} e^{i\theta k} g(k)$ is defined by a value $V \in \mathbb{C}$ under $\mathfrak{T}$. Then, we have:*

$$\sum_{k=0}^{n-1} e^{i\theta k} g(k) = \sum_{k=0}^{\infty} e^{i\theta k} g(k) - \sum_{k=n}^{\infty} e^{i\theta k} g(k) \tag{5.3.2}$$

$$\frac{d}{dn} \sum_{k=0}^{n-1} e^{i\theta k} g(k) = \sum_{k=0}^{n-1} \frac{d}{dk} \left(e^{i\theta k} g(k) \right) - \sum_{k=0}^{\infty} \frac{d}{dk} \left(e^{i\theta k} g(k) \right) \tag{5.3.3}$$

Proof Equation 5.3.2 was already established in Theorem 5.3. To prove Eq. 5.3.3, we first note by Theorem 5.3 that:

$$f_G(n) = \sum_{r=0}^{\infty} \frac{\Theta_r}{r!} \left[g^{(r)}(0) - e^{i\theta n} g^{(r)}(n) \right] \qquad (5.3.4)$$

In Theorem 5.3, we have also shown that $\sum_{k=0}^{\infty} e^{i\theta k} g(k)$, interpreted using $\mathfrak{T}$, is formally given by:

$$\sum_{k=0}^{\infty} e^{i\theta k} g(k) = \sum_{k=0}^{n-1} e^{i\theta k} g(k) + \sum_{r=0}^{\infty} \frac{\Theta_r}{r!} e^{i\theta n} g^{(r)}(n) \qquad (5.3.5)$$

Now, using the differentiation rule of simple finite sums, we have:

$$\frac{d}{dn} \sum_{k=0}^{n-1} e^{i\theta k} g(k) = \sum_{k=0}^{n-1} \frac{d}{dk} \left(e^{i\theta k} g(k) \right) + c \qquad (5.3.6)$$

On the other hand, differentiating both sides of Eq. 5.3.4 formally yields:

$$f_G'(n) = -\sum_{r=0}^{\infty} \frac{\Theta_r}{r!} \frac{d}{dn} \left(e^{i\theta n} g^{(r)}(n) \right) \qquad (5.3.7)$$

Equating Eqs. 5.3.6 and 5.3.7 yields:

$$c = -\sum_{r=0}^{\infty} \frac{\Theta_r}{r!} \frac{d}{dn} \left(e^{i\theta n} g^{(r)}(n) \right) - \sum_{k=0}^{n-1} \frac{d}{dk} \left(e^{i\theta k} g(k) \right) \qquad (5.3.8)$$

Comparing last equation with Eq. 5.3.5 implies that:

$$c = -\sum_{k=0}^{\infty} \frac{d}{dk} \left(e^{i\theta k} g(k) \right) \qquad (5.3.9)$$

Plugging Eq. 5.3.9 into Eq. 5.3.6 yields the desired result. $\qquad\qquad\square$

Lemma 5.4 *Suppose we have a simple finite sum of the form* $f(n) = \sum_{k=0}^{n-1} e^{i\theta k} g(k)$, *where the infinite sum* $\sum_{k=0}^{\infty} e^{i\theta k} g(k)$ *is well-defined in* $\mathfrak{T}$. *Then, the unique natural generalization* $f_G(n)$ *that is consistent with the successive polynomial approximation method of Theorem 2.1 is formally given by the series expansion:*

$$f_G(n) = \sum_{r=1}^{\infty} \frac{c_r}{r!} n^r, \text{ where } c_r = -\sum_{k=0}^{\infty} \frac{d^r}{dk^r} \left(e^{i\theta k} g(k) \right) \qquad (5.3.10)$$

Here, the infinite sums c_r *are interpreted using the generalized definition* $\mathfrak{T}$.

Proof Follows immediately from Lemma 5.3. $\qquad\qquad\square$

Note that in Lemma 5.4, the function $f(n)$ behaves as if it were a convergent function with respect to the rules of infinitesimal calculus as discussed earlier. This result can be generalized to arbitrary oscillating sums of the form $\sum_{k=0}^{n-1} s_k\, g(k)$, in which s_k is a periodic sign sequence with period p that satisfies the condition $\sum_{k=0}^{p-1} s_k = 0$, by decomposing s_k into multiple sequences of the form $e^{i\theta k}$ using the Discrete Fourier Transform (DFT) as shown earlier in Lemma 5.2. In fact, it can be generalized even further as the following theorem shows.

Theorem 5.5 *Given a simple finite sum $\sum_{k=0}^{n-1} g(k)$, where $\sum_{k=0}^{\infty} g(k)$ exists in $\mathfrak{T}$, then we have:*

$$f_G(n) = \sum_{k=0}^{\infty} g(k) - \sum_{k=n}^{\infty} g(k) \tag{5.3.11}$$

$$f_G^{(r)}(n) = \sum_{k=0}^{n-1} g^{(r)}(k) - \sum_{k=0}^{\infty} g^{(r)}(k), \qquad r \geq 1 \tag{5.3.12}$$

Similarly the indefinite integral is given by:

$$\int^n f_G(t)\, dt = \sum_{k=0}^{n-1} \int^k g(t)\, dt + n \sum_{k=0}^{\infty} g(k) + c \tag{5.3.13}$$

Here, c is an arbitrary constant of integration, and all infinite sums are interpreted using the generalized definition $\mathfrak{T}$.

Proof We can rewrite $f(n)$ as $f(n) = \sum_{k=0}^{n-1} e^{i\theta k}(e^{-i\theta k} g(k))$ and use the previous results.   □

Exercise 5.9 In Exercise 2.3, it was shown that:

$$\sum_{k=0}^{n-1} \sin k = \alpha \sin n + \beta(1 - \cos n), \tag{5.3.14}$$

for some constants α and β. Compute the values of α and β by solving a system of linear equations, which correspond to setting $n = 1$ and $n = 2$ in the above equation. Then, prove that $\beta = \sum_{k=0}^{\infty} \sin k$ by taking the $\mathfrak{T}$ sequence limit of both sides in the above equation (Hint: show that the $\mathfrak{T}$ sequence limit of $\sin n$ and $\cos n$ is zero). Finally, verify numerically that $\beta = \sum_{k=0}^{\infty} \sin k$ indeed holds using the summability method χ.

Exercise 5.10 Use the identity $\sum_{k=0}^{n-1} \sin(\pi k/n) = \cot(\pi/(2n))$ and the rules of summability calculus for oscillating composite sums to derive the following identity:

$$\sum_{k=0}^{n-1} \left(1 - \frac{k}{n}\right) \cos \frac{\pi k}{n} = \frac{1}{2}\left(1 + \frac{1}{n} \csc^2 \frac{\pi}{2n}\right) \tag{5.3.15}$$

(Hint: use the differentiation rule and the fact that for any θ, the series $\sum_{k=0}^{\infty} \cos(k\theta)$ is defined under $\mathfrak{T}$ by the value $\frac{1}{2}$.)

Exercise 5.11 Prove that $\sum_{k=0}^{n-1} \cos(\pi \frac{2k+1}{2n+1}) = \frac{1}{2}$ (Hint: consider the well-known trigonometric identity for the product $\cos x \cdot \sin y$). Use this fact and the results of this chapter to prove that:

$$\sum_{k=0}^{\infty} \sin\left(\pi \frac{2k+1}{2n+1}\right) = \sum_{k=0}^{n-1} \sin\left(\pi \frac{2k+1}{2n+1}\right) \cdot \left(1 - \frac{2k+1}{2n+1}\right) \tag{5.3.16}$$

The same technique can be applied to express other summable divergent series as composite finite sums.

Exercise 5.12 Prove that:

$$\sum_{k=0}^{n-1} k \cos \frac{2\pi k}{n} = -\frac{n}{2}$$

(Hint: begin by writing $f(n) = \sum_{k=0}^{n-1} \cos \frac{2\pi k}{n}$, make use of the fact that $f(n) = 0$, and differentiate using the chain rule and Theorem 5.3).

Exercise 5.13 Show that the following function:

$$f(n) = \frac{1}{2}\left(\sum_{k=0}^{n-1} g(k) + (-1)^n \sum_{k=0}^{n-1}(-1)^{-k} g(k)\right) + \frac{1}{2}\left[(f_0 + f_1) + (-1)^n (f_0 - f_1)\right] \tag{5.3.17}$$

satisfies the recurrence identity $f(n) = g(n-2) + f(n-2)$ and the initial conditions $f(0) = f_0$ and $f(1) = f_1$. Consider the following example:

$$h(n) = \int_0^{\infty} t^s e^{-t^2} dt, \tag{5.3.18}$$

and write $f(n) = \log h(n)$. Prove that $f(n) = \log \frac{n-1}{2} + f(n-2)$. Use this fact and the previous result to prove that:

$$\int_0^{\infty} \log t\, e^{-t^2} dt = -\frac{\sqrt{\pi}}{4} (\gamma + 2\log 2), \tag{5.3.19}$$

which is also mentioned in [Seb02]. Use the fact that $h(n)$ is real-valued when $n \in \mathbb{R}$ and trigonometric identities to prove that for all $n \in \mathbb{R}$, we have:

$$\sum_{k=0}^{n-1} \sin(\pi(n-k)) \log(1+k) = \sin(\pi n) \log \sqrt{\frac{2}{\pi}} \qquad (5.3.20)$$

Moreover, we also have:

$$\sum_{k=0}^{\infty} (-1)^k \log(1+k+n) = 2 \log \left(\frac{n}{2}\right)! - \log n! + n \log 2 - \log \sqrt{\frac{\pi}{2}}$$

Setting $n = \frac{1}{2}$, we conclude that:

$$\log(1/4)! = \frac{1}{2} \log \frac{\pi}{4\sqrt{2}} - \frac{1}{2} \sum_{k=0}^{\infty} (-1)^k \log(1+2k),$$

which gives a close-form expression for the summable divergent series $\sum_{k=0}^{\infty} (-1)^k \log(1 + 2k)$. Verify this result numerically using the summability method χ.

5.4 Summary

In this chapter, we used the generalized definition of series $\mathfrak{T}$ given earlier in Chap. 4 to derive the analogs of the Euler-Maclaurin summation formula for oscillating sums. These results were used to deduce the asymptotic expressions of oscillating sums, accelerate series convergence, and even derive the exact analytic expressions of summable divergent series in some occasions. We have also shown the remarkable fact that, as far as the foundational rules of summability calculus are concerned, summable divergent series behave as if they were convergent.

References

[Ber85] B.C. Berndt, Ramanujan's theory of divergent series, in *Ramanujan's Notebooks* (Springer, Berlin, 1985)
[BS75] B. Berndt, L. Schoenfeld, Periodic analogues of the Euler-Maclaurin and Poisson summation formulas with applications to number theory. Acta Arith. **28**(1), 23–68 (1975)
[BBD89] J.M. Borwein, P.B. Borwein, K. Dilcher, Pi, Euler numbers, and asymptotic expansions. Am. Math. Mon. **96**(8), 681–687 (1989)
[BCM09] J.M. Borwein, N.J. Calkin, D. Manna, Euler-Boole summation revisited. Am. Math. Mon. **116**(5), 387–412 (2009)

[CS00] J. Choi, H.M. Srivastava, Certain classes of series associated with the zeta function and multiple gamma functions. J. Comput. Appl. Math. **118**, 87–109 (2000)

[CVZ00] H. Cohen, F.R. Villegas, D. Zagier, Convergence acceleration of alternating series. Exp. Math. **9**, 3–12 (2000)

[Har49] G.H. Hardy, *Divergent Series* (Oxford University Press, New York, 1949)

[Jor65] K. Jordán, *Calculus of Finite Differences* (Chelsea Publishing Company, Vermont, 1965)

[Rob55] H. Robbins, A remark on Stirling's formula. Am. Math. Mon. **62**, 26–29 (1955)

[San06] E. Sandifer, How Euler did it: divergent series, June 2006. Available at The Mathematical Association of America (MAA). http://www.maa.org/news/howeulerdidit.html

[Seb02] P. Sebha, Collection of formulae for Euler's constant, 2002. http://scipp.ucsc.edu/~haber/archives/physics116A06/euler2.ps, Retrieved on March 2011

Chapter 6
Computing Finite Sums

Check for updates

> *The future influences the present, just as much as the past.*
> F. Nietzsche (1844–1900)

Abstract In this chapter, we derive methods for computing the values of fractional finite sums. The motivation behind the methods developed in this chapter is twofold. First, the Euler-Maclaurin summation formula and all of its analogs diverge too rapidly and, hence, they cannot be used to calculate finite sums, not even with the help of summability methods. Second, computing the Taylor series expansions and using them, in turn, to calculate fractional finite sums by rigidity is a tedious task and is far from being a satisfactory approach. In this chapter, we resolve those limitations by providing a simple direct method of computing fractional finite sums. We show how well-known historical results, such as Euler's infinite product formula for the gamma function, fall as particular cases of these more general results. We also show why the Euler-Maclaurin summation formula is connected to polynomial approximation. Finally, we extend results to composite finite sums.

In this chapter, we derive methods for computing the values of fractional finite sums $\sum_{k=0}^{n-1} s_k\, g(k, n)$ where s_k is an arbitrary periodic sign sequence (see Chap. 5) and $n \in \mathbb{C}$. The motivation behind the methods developed in this chapter is twofold. First, all of the analogs of the Euler-Maclaurin summation formula diverge too rapidly and, hence, they cannot be used to *calculate* finite sums, not even with the help of summability methods. Second, computing the Taylor series expansions and using them, in turn, to calculate fractional finite sums by rigidity is a tedious task and is far from being a satisfactory approach. In this chapter, we resolve those limitations by providing a simple direct method of computing $\sum_{k=0}^{n-1} s_k\, g(k, n)$ for all $n \in \mathbb{C}$.

In the case of simple finite sums $f(n) = \sum_{k=0}^{n-1} g(k)$ in which $g(n)$ is of a finite polynomial order m, the fairly recent work of Müller and Schleicher [MS10] has captured the general principle. Here, and as discussed earlier in Chap. 1, the basic idea is to evaluate the finite sum asymptotically using approximating polynomials and to propagate results backwards using the recurrence $f(n) = g(n-1) + f(n-1)$.

© Springer International Publishing AG 2018
I. M. Alabdulmohsin, *Summability Calculus*,
https://doi.org/10.1007/978-3-319-74648-7_6

To be more specific, the Müller-Schleicher method can be described as follows. Suppose that for a fixed function $g(t)$, there exists a family of polynomials $P_N(t)$ such that for any fixed $0 < W < \infty$, the following condition holds:

$$\lim_{N \to \infty} \int_{N-W}^{N+W} \left| g(t) - P_N(t) \right| dt = 0 \qquad (6.0.1)$$

Equation 6.0.1 states that the two functions $g(k)$ and $P_N(k)$ are nearly indistinguishable from each other at the vicinity of N if N is sufficiently large. If this condition holds, then for any fixed $n \in \mathbb{N}$ and $N \gg 1$, we have:

$$\sum_{k=N}^{n+N} g(k) \approx \sum_{k=N}^{n+N} P_N(k)$$

However, the second finite sum can be evaluated using the Bernoulli-Faulhaber formula for all $n \in \mathbb{C}$. Consequently, we can use the finite sum on the right-hand side of the above equation *as a definition* for the finite sum on the left-hand side when N is sufficiently large. Of course, this is merely an approximation. Nonetheless, we can make it *exact* for all $n \in \mathbb{C}$ by taking the limit as $N \to \infty$ and using the two defining properties of finite sums, which yield:

$$\sum_{k=0}^{n-1} g(k) = \lim_{N \to \infty} \left\{ \sum_{k=N}^{n+N} g(k) - \sum_{k=n}^{n+N} g(k) + \sum_{k=0}^{N-1} g(k) \right\} \qquad (6.0.2)$$

Now, choosing $N \in \mathbb{N}$ allows us to evaluate the sums $\sum_{k=n}^{n+N} g(k)$ and $\sum_{k=0}^{N-1} g(k)$ directly by definition. Additionally, taking $N \to \infty$ allows us to evaluate the *fractional* sum $\sum_{k=N}^{n+N} g(k)$ with an arbitrary accuracy using the polynomial approximation method discussed earlier. Consequently, we can evaluate the original fractional sum $\sum_{k=0}^{n-1} g(k)$ with an arbitrary degree of accuracy by applying the polynomial approximation method above at the limit $N \to \infty$.

Obviously, such an approach is not restricted to analytic functions as pointed out by Müller and Schleicher in their original work. For instance, if we define $f(n) = \sum_{k=1}^{n} 1/\lceil k \rceil$, where $\lceil x \rceil$ is the *ceiling* function, then the natural generalization implied by the above method is given by the *discrete harmonic function* $f(n) = H_{\lceil n \rceil}$. This is the essence of the Müller-Schleicher method.

However, if $g(k)$ is analytic and has a finite polynomial order m, then we can use the Taylor theorem to conclude that $g(k)$ is approximable by a polynomial with degree m in the sense specified by Eq. 6.0.1. This can be easily established using, for instance, the Lagrange reminder form.

For example, suppose we would like to analyze the finite sum $f(n) = \sum_{k=0}^{n-1} (1 + k) \log (1 + k)$, which is the log-hyperfactorial function [Weib]. The Taylor series expansion for the function $g(k)$ around a point $k = N$ is given by:

$$(1+k) \log (1 + k) = (1+N) \log (1 + N) + (1 + \log (1 + N)) (k-N) + \frac{1}{2 (\kappa + 1)} (k-N)^2$$

Here, κ is some constant in the interval $[k, N]$. Now, suppose $|k - N| < W$ for some fixed W and $N \to \infty$, then the last term vanishes, and we obtain:

$$(1 + k) \log (1 + k) \sim (1 + N) \log (1 + N) + (1 + \log (1 + N)) (k - N) \quad (6.0.3)$$

Therefore, the iterated function $(1 + k) \log (1 + k)$ becomes affine at the specified limit. Having this in mind, one can now evaluate the finite sum $\sum_{k=0}^{n-1} (1 + k) \log (1 + k)$ using Eqs. 6.0.3 and 6.0.2 for fractional values of n. As illustrated above, the Müller-Schleicher method is quite rich but it can only be applied to the case of simple finite sums for which the iterated function $g(k)$ is asymptotically of a finite polynomial order.

In this section, we show that the aforementioned approach is a particular case of a more general formal method. Here, we present the general statement that is applicable to all simple finite sums, even those in which $g(k)$ is not asymptotically of a finite polynomial order, composite finite sums of the form $f(n) = \sum_{k=0}^{n-1} g(k, n)$ and even oscillating composite finite sums of the form $\sum_{k=0}^{n-1} s_k \, g(k, n)$ for some arbitrary periodic sign sequence s_k. Besides, we also demonstrate that the general formal method can be used to *accelerate* the speed of convergence. Finally, we will establish the earlier claim that the Euler-Maclaurin summation formula corresponds to polynomial fitting.

6.1 Evaluating Semi-linear Simple Finite Sums

Similar to the approach of Chap. 2, we will begin our treatment with the case of semi-linear simple finite sums and generalize results afterward.

6.1.1 Main Results

The case of semi-linear simple finite sums is summarized in the following theorem.

Theorem 6.1 (Müller-Schleicher [MS10]) *For a given* semi-linear *simple finite sum of the form* $f(n) = \sum_{k=0}^{n-1} g(k)$ *, let* $f_G(n)$ *denote the unique natural generalization implied by the successive polynomial approximation method of Theorem 2.1. Then,* $f_G(n)$ *can be evaluated using the following expression:*

$$f_G(n) = \lim_{s \to \infty} \left\{ n \, g(s) + \sum_{k=0}^{s-1} g(k) - g(k + n) \right\} \quad (6.1.1)$$

Proof The proof consists of three parts. First, we need to show that the initial condition holds. Second, we need to show that the required recurrence identity also holds. Third, we need to show that the function $f_G(n)$ given by Eq. 6.1.1 is equivalent to the unique natural generalization implied by summability calculus.

The proof that the initial condition holds is straightforward. Plugging $n = 0$ into Eq. 6.1.1 yields the empty sum rule $f(0) = 0$. To show that the recurrence identity holds, we note that:

$$f_G(n) - f_G(n-1) = \lim_{s \to \infty} \{g(s) + g(n-1) - g(s+n-1)\} \qquad (6.1.2)$$

However, because $g(n)$ is nearly-convergent by assumption, we have:

$$\lim_{s \to \infty} \{g(s) - g(s+n-1)\} = 0, \text{ for all } n \in \mathbb{C} \qquad (6.1.3)$$

Thus, the function $f_G(n)$ given by Eq. 6.1.1 indeed satisfies the recurrence identity:

$$f_G(n) - f_G(n-1) = g(n-1) \qquad (6.1.4)$$

Finally, to show that $f_G(n)$ is formally identical to the unique natural generalization implied by summability calculus, we differentiate both sides of Eq. 6.1.1, which yields:

$$f_G'(n) = \lim_{s \to \infty} \left\{ g(s) - \sum_{k=0}^{s-1} g'(k+n) \right\} = \sum_{k=0}^{n-1} g'(k) + \lim_{s \to \infty} \left\{ g(s) - \sum_{k=0}^{s+n-1} g'(k) \right\}$$

$$= \sum_{k=0}^{n-1} g'(k) + \lim_{s \to \infty} \left\{ g(s) - g(s+n) + g(s+n) - \sum_{k=0}^{s+n-1} g'(k) \right\}$$

$$= \sum_{k=0}^{n-1} g'(k) + \lim_{s \to \infty} \{ g(s) - g(s+n-1) \} + \lim_{s \to \infty} \left\{ g(s+n-1) - \sum_{k=0}^{s+n-1} g'(k) \right\}$$

As mentioned earlier, since g is nearly convergent, we have $\lim_{s \to \infty} \{g(s) - g(s + n - 1)\} = 0$ for any fixed $n \in \mathbb{C}$. Therefore:

$$f_G'(n) = \sum_{k=0}^{n-1} g'(k) + \lim_{N \to \infty} \left\{ g(N-1) - \sum_{k=0}^{N-1} g'(k) \right\} \qquad (6.1.5)$$

However, the last equation is identical to Theorem 2.2. Repeating the same process for all higher-order derivatives shows that the function $f_G(n)$ given by Eq. 6.1.1 is indeed the unique natural generalization to the simple finite sum. $\qquad \square$

Corollary 6.1 *If $g(n) \to 0$ as $n \to \infty$, then the unique natural generalization to the simple finite sum $f(n) = \sum_{k=0}^{n-1} g(k)$ for all $n \in \mathbb{C}$ is given by:*

$$f_G(n) = \sum_{k=0}^{\infty} \left[g(k) - g(k+n) \right] \qquad (6.1.6)$$

Proof Follows immediately by Theorem 6.1.   □

As discussed earlier, because a simple finite sum is asymptotically linear in any bounded region W, we can push W to infinity such that the simple finite sum becomes "exactly" linear. In this case, we know that $\sum_{k=s}^{s+W-1} g(k) \to W g(s)$ for any fixed $0 \le W < \infty$. This provides a convenient method of evaluating fractional sums. To evaluate simple finite sums of the form $\sum_{k=0}^{n-1} g(k)$, for $n \in \mathbb{C}$, we use the backward recurrence relation $f(n-1) = f(n) - g(n-1)$. Theorem 6.1 shows that such an approach yields the unique natural generalization to simple finite sums as implied by the successive polynomial approximation method of Theorem 2.1.

We illustrate the statement of Theorem 6.1 next with a few examples.

6.1.2 Examples

6.1.2.1 The Factorial Function and Euler's Infinite Product Formula

First, we will start with the log-factorial function $\varpi(n) = \sum_{k=0}^{n-1} \log(1+k)$. Because $\log(1+k)$ is nearly-convergent, we use Theorem 6.1, which yields:

$$\log n! = \lim_{s \to \infty} \left\{ n \log(1+s) + \sum_{k=0}^{s-1} \log\left(\frac{k+1}{k+n+1}\right) \right\}$$

$$= \lim_{s \to \infty} \left\{ n \sum_{k=1}^{s} \log\left(1 + \frac{1}{k}\right) + \sum_{k=0}^{s} \log\left(\frac{k+1}{k+n+1}\right) \right\}$$

$$= -\log(1+n) + \sum_{k=1}^{\infty} \log\left(\left(1 + \frac{1}{k}\right)^n \frac{k+1}{k+n+1}\right)$$

Therefore, we have:

$$n! = \prod_{k=1}^{\infty} \left(1 + \frac{1}{k}\right)^n \frac{k}{k+n} \tag{6.1.7}$$

Equation 6.1.7 is the famous infinite product formula for the factorial function derived by Euler. In addition, we know by Theorem 6.1 that Eq. 6.1.7 is an alternative definition of $\Gamma(n+1)$, where Γ is the gamma function because the log-gamma function is the unique natural generalization of the log-factorial function as proved earlier in Proposition 2.4.

6.1.2.2 The Sum of Square Roots Function

Our second example is the sum of the square roots function $\sum_{k=0}^{n-1} \sqrt{(1+k)}$. Previously, we derived its series expansion in Eq. 2.5.5 whose radius of convergence was $|n| \leq 1$. In Chap. 4, we used the summability method χ to evaluate the Taylor series expansion in a larger region over the complex plane $\mathbb{C}$. Now, we use Theorem 6.1 to evaluate the function for all $n \in \mathbb{C}$ directly.

Using Theorem 6.1, the sum of square roots function can be rewritten as:

$$\sum_{k=0}^{n-1} \sqrt{(1+k)} = \lim_{s \to \infty} \left\{ n\sqrt{1+s} + \sum_{k=0}^{s-1} (\sqrt{k+1} - \sqrt{k+n+1}) \right\}$$

$$= \sum_{k=0}^{\infty} (n+1)\sqrt{k+1} - n\sqrt{k} - \sqrt{k+n+1} \qquad (6.1.8)$$

As stated in Theorem 6.1, this definition holds for fractional values of n as well. The function in Eq. 6.1.8 is plotted in Fig. 6.1. The points highlighted with markers in the figure are evaluated using the original discrete definition of the simple finite sum. Clearly, the function that results from applying Theorem 6.1 to the sum of square roots function $\sum_{k=0}^{n-1} \sqrt{k+1}$ correctly interpolates the discrete points as expected.

Remark 6.1 Note that the value of $f(n) = \sum_{k=0}^{n-1} \sqrt{k+1}$ at $n = -1$ is zero as shown in Fig. 6.1, which follows directly from Eq. 6.1.8. This is not a coincidence. If we write $h(n) = \sum_{k=0}^{n-1} \sqrt{k}$, we note that $h(n) = f(n-1)$. However, $h(0) = 0$ by the empty sum rule. So, we must have $f(-1) = 0$ as well (see Exercise 2.13).

Exercise 6.1 Use Theorem 6.1 to prove that the harmonic numbers $H_n = \sum_{k=0}^{n-1} \frac{1}{1+k}$ can be computed for any $n \in \mathbb{C}$ using the series:

$$H_n = \sum_{k=1}^{\infty} \frac{n}{k(k+n)} \qquad (6.1.9)$$

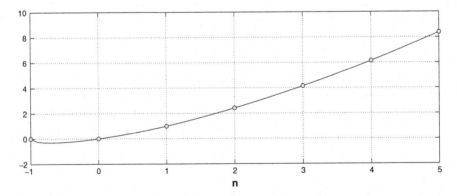

Fig. 6.1 The function $\sum_{k=1}^{n} \sqrt{k}$ evaluated using Theorem 6.1. Values highlighted using markers are exact and are evaluated directly using the original discrete definition

Exercise 6.2 Show that if the series $\sum_{k=0}^{\infty} g(k)$ converges, Theorem 6.1 implies that $\sum_{k=0}^{n-1} g(k) = \sum_{k=0}^{\infty} g(k) - \sum_{k=n}^{\infty} g(k)$.

Exercise 6.3 Use Theorem 6.1 and the integration rule of simple finite sums to derive the Weierstrass product [Spo94]:

$$n! = e^{-\gamma n} \prod_{k=1}^{\infty} \left(1 + \frac{n}{k}\right)^{-1} e^{\frac{n}{k}} \tag{6.1.10}$$

(Hint: start with the harmonic numbers and integrate accordingly)

Exercise 6.4 Use Theorem 6.1 to prove that $H_{1/2} = 2(1 - \log 2)$, where H_n is the nth harmonic number.

6.2 Evaluating Arbitrary Simple Finite Sums

6.2.1 Main Results

To evaluate arbitrary simple finite sums, we extend the previous results using the following theorem.

Theorem 6.2 *Given a simple finite sum of the form* $f(n) = \sum_{k=0}^{n-1} g(k)$, *let* $f_G(n)$ *denote the unique natural generalization implied by the successive polynomial approximation method of Theorem 2.1. Then,* $f_G(n)$ *for all* $n \in \mathbb{C}$ *is formally given by:*

$$f_G(n) = \sum_{r=0}^{\infty} \frac{b_r(n)}{r!} g^{(r)}(s) + \sum_{k=0}^{s-1} g(k) - g(k+n) \tag{6.2.1}$$

Here, Eq. 6.2.1 holds formally *for any value* s, *where* $b_r(n) = \sum_{k=0}^{n-1} k^r$, *i.e. a polynomial given by the Bernoulli-Faulhaber formula.*

Moreover, if $g(s)$ *has a finite polynomial order* m, *then* $f_G(n)$ *can be computed using the following limit:*

$$f_G(n) = \lim_{s \to \infty} \left\{ \sum_{r=0}^{m} \frac{b_r(n)}{r!} g^{(r)}(s) + \sum_{k=0}^{s-1} g(k) - g(k+n) \right\} \tag{6.2.2}$$

Proof Again, the proof consists of three parts.

First, plugging $n = 0$ into Eq. 6.2.1 yields the empty sum rule $f_G(0) = 0$. Second, to show that the recurrence identity holds, we start from Eq. 6.2.1:

$$f_G(n) - f_G(n-1) = \left(\sum_{r=0}^{\infty} \frac{g^{(r)}(s)}{r!} (n-1)^r \right) + g(n-1) - g(n+s-1) \tag{6.2.3}$$

The first term is a Taylor series expansion and it is formally equal to $g(n + s - 1)$. Consequently, we indeed have:

$$f_G(n) - f_G(n - 1) = g(n - 1) \qquad (6.2.4)$$

So, the recurrence identity holds formally. Finally, to show that the function $f_G(n)$ given by Eq. 6.2.1 is identical to the unique natural generalization implied by summability calculus, we differentiate both sides of Eq. 6.2.1, which yields:

$$f_G'(n) = \sum_{r=0}^{\infty} \frac{b_r'(n)}{r!} g^{(r)}(s) - \sum_{k=0}^{s-1} g'(k + n) \qquad (6.2.5)$$

Upon rearranging the terms, we have:

$$f_G'(n) = \sum_{k=0}^{n-1} g'(k) + \left[\sum_{r=0}^{\infty} \frac{b_r'(n)}{r!} g^{(r)}(s) - \sum_{k=0}^{s+n-1} g'(k) \right] \qquad (6.2.6)$$

Now, if we note that the *Bernoulli polynomials* are defined by [DM09]:

$$B_r(x) = \sum_{k=0}^{r} \binom{r}{k} B_k \, n^{r-k} \qquad (6.2.7)$$

We observe that $b_r'(n) = B_r(n)$. So:

$$f_G'(n) = \sum_{k=0}^{n-1} g'(k) + \left[\sum_{r=0}^{\infty} \frac{B_r(-n)}{r!} g^{(r)}(s) - \sum_{k=0}^{s+n-1} g'(k) \right] \qquad (6.2.8)$$

Now, we use Eq. 2.4.1 to rewrite the above equation as:

$$f_G'(n) = \sum_{k=0}^{n-1} g'(k) - \sum_{k=0}^{s+n-1} g'(k) + \sum_{r=0}^{\infty} \frac{n^r}{r!} \frac{d}{ds} \sum_{k=0}^{s-1} g^{(r)}(k)$$

$$= \sum_{k=0}^{n-1} g'(k) - \sum_{k=0}^{s+n-1} g'(k) + \frac{d}{ds} \sum_{k=0}^{s-1} \sum_{r=0}^{\infty} \frac{n^r}{r!} g^{(r)}(k)$$

$$= \sum_{k=0}^{n-1} g'(k) - \sum_{k=0}^{s+n-1} g'(k) + \frac{d}{ds} \sum_{k=0}^{s-1} g(k + n)$$

$$= \sum_{k=0}^{n-1} g'(k) - \sum_{k=0}^{s+n-1} g'(k) + \frac{d}{ds} \sum_{k=0}^{s+n-1} g(k)$$

Here, we used the translation-invariance property, which is derived from the Euler-Maclaurin summation formula. Finally, we note from the differentiation rule of simple finite sums that the quantity $\left[-\sum_{k=0}^{s+n-1} g'(k) + \frac{d}{ds}\sum_{k=0}^{s+n-1} g(k)\right]$ is a constant that is independent of s and n. Consequently, the quantity $\left[\sum_{r=0}^{\infty} \frac{B_r(-n)}{r!} g^{(r)}(s) - \sum_{k=0}^{s+n-1} g'(k)\right]$ in Eq. 6.2.8 must also be a constant that is formally independent of both s and n. So, we can choose any values of these two variables to determine the value of that constant. If we choose $n = 0$, Eq. 6.2.8 is rewritten as:

$$f'_G(n) = \sum_{k=0}^{n-1} g'(k) + \left[\sum_{r=0}^{\infty} \frac{B_r}{r!} g^{(r)}(s) - \sum_{k=0}^{s-1} g'(k)\right] \tag{6.2.9}$$

Here, B_r are Bernoulli numbers. However, Eq. 6.2.9 is identical to Theorem 2.3, which was shown to be implied by the successive polynomial approximation method of Theorem 2.1. Therefore, $f_G(n)$ given by Eq. 6.2.1 is indeed the unique natural generalization implied by summability calculus.   □

6.2.2 Examples

We illustrate Theorem 6.2 with a few examples next.

6.2.2.1 The Superfactorial Function

In our first example, we apply Theorem 6.2 on the superfactorial function [Weia] $S(n) = \prod_{k=1}^{n} k!$. After simplification, this yields Eq. 6.2.10.

$$\prod_{k=1}^{n} k! = \prod_{k=1}^{\infty} \left(1 + \frac{1}{k}\right)^{n(n+1)/2} k^n \frac{k!}{\Gamma(k+n+1)} \tag{6.2.10}$$

One immediate consequence of Eq. 6.2.10 is given by:

$$\frac{S'(0)}{S(0)} = \sum_{k=1}^{\infty} \left(\frac{\log k + \log(k+1)}{2} + \gamma - H_k\right) \tag{6.2.11}$$

Using the series expansion for the log-superfactorial function that was derived earlier in Eq. 2.5.17, we have:

$$\sum_{k=1}^{\infty} \left(\frac{\log k + \log(k+1)}{2} + \gamma - H_k\right) = \frac{\log(2\pi) - 1}{2} - \gamma \tag{6.2.12}$$

Equation 6.2.12 shows that the quantity $H_n - \gamma$ converges to the average $(\log k + \log (k + 1))/2$ faster than its convergence to $\log n$. This is clearly demonstrated by the fact that the sum of the approximation errors given by the earlier equation converges whereas the sum $\sum_{k=1}^{\infty} (\log k + \gamma - H_k)$ diverges. Note that, by contrast, we have Eq. 5.1.33.

6.2.2.2 Accelerating the Speed of Convergence

If $g(n)$ is of a finite polynomial order $m < \infty$, we can use Eq. 6.2.2. However, adding more terms to the sum $\sum_{r=0}^{m} \frac{b_r(n)}{r!} g^{(r)}(s)$ improves the speed of convergence. For example, if we return to the factorial function and choose $m = 1$, we obtain:

$$\log n! = \lim_{s \to \infty} \left\{ n \log (1 + s) + \frac{n^2 - n}{2(1 + s)} + \sum_{k=1}^{s} \log \frac{k}{k + n} \right\} \qquad (6.2.13)$$

By contrast, Euler's infinite product formula omits the term $\frac{n^2-n}{2(1+s)}$. So, if we wish to compute a value such as $(1/2)!$, and choose $s = 10^4$, we obtain numerically the two estimates $(1/2)! \approx 0.8862380$ and $(1/2)! \approx 0.8862269$, which correspond to choosing $m = 0$ and $m = 1$ respectively. Here, the approximation error is 1×10^{-5} if $m = 0$ but it is less than 2×10^{-14} if we use $m = 1$. Hence, the formal expression in Theorem 6.2 can to be used as a method of improving the speed of convergence.

6.2.3 The Euler-Maclaurin Summation Formula Revisited

Before we conclude this section, it is worth pointing out that Theorem 6.2 indicates that the Euler-Maclaurin summation formula is very closely related to Taylor's theorem. In fact, the expression in Eq. 6.2.1 resembles Taylor series expansions, and we do obtain a Taylor series by simply subtracting $f(n - 1)$ from $f(n)$. In fact, the Euler-Maclaurin summation formula and the Taylor series expansion can both be derived using very similar approaches [Lam01]. It should not be surprising, therefore, to note that analytic summability methods that are consistent with the $\mathfrak{T}$ definition of series can work reasonably well even with the Euler-Maclaurin summation formula. We have already illustrated this in Table 4.1, in which we showed that the series $\sum_{r=0}^{\infty} B_r$ is "almost" summable using χ. In particular, using χ with a small value of n yields reasonably accurate figures, and accuracy improves as we increase n until we reach about $n = 35$, after which the error begins to increase. This is also true for the series $1 + \sum_{r=1}^{\infty} \frac{B_r}{r}$, which is formally equivalent to the Euler constant γ as discussed earlier.

Such a close correspondence between the Euler-Maclaurin summation formula and the Taylor series expansion also shows that the Euler-Maclaurin formula is an asymptotic expression. Here, if $g(n)$ is of a finite polynomial order m, then by

Taylor's theorem:

$$\lim_{n\to\infty} \left\{ g(n+h) - g(n) - \sum_{r=1}^{m} \frac{g^{(r)}(n)}{r!} h^r \right\} = 0 \tag{6.2.14}$$

Equation 6.2.14 is easily established upon using, for example, the Lagrange remainder form. From this result, we conclude that if $g(n)$ is of a finite polynomial order m, then:

$$\lim_{n\to\infty} \left\{ \frac{d}{dn} \sum_{k=a}^{n} g(k) - \sum_{r=0}^{m} \frac{B_r}{r!} g^{(r)}(n) \right\} = 0 \tag{6.2.15}$$

The last equation was the basis of many identities derived in Sect. 2.4. This is a simple explanation for why the Euler-Maclaurin summation formula is more correctly interpreted as an asymptotic series expansion.

Finally, in Chap. 2, we argued that summability calculus yields the unique most natural generalization for semi-linear simple finite sums by reducing the argument for unique natural generalization to the case of *linear* fitting since all semi-linear simple finite sums are asymptotically linear in any bounded region. Here, we can see that for the general case of simple finite sums, summability calculus yields a unique natural generalization as well because it implicitly relies on polynomial fitting. This was previously established using the successive polynomial approximation method in Theorem 2.1 but we can see this fact, again, in Theorem 6.2. So, where does polynomial fitting show up in Theorem 6.2?

To answer the latter question, we return to the proof of Theorem 6.2. In Eq. 6.2.1, we note that the crucial property of the functions $b_r(n)$ that makes $f_G(n)$ satisfy the initial condition and the required recurrence identity is the condition $b_r(n) - b_r(n-1) = (n-1)^r$. However, it is obvious that an infinite number of functions $b_r(n)$ exist that can satisfy this condition. Nevertheless, because this condition happens to be satisfied by the Bernoulli-Faulhaber formula, which is the *unique* polynomial solution to the required identity, defining $b_r(n) = \sum_{k=0}^{n-1} k^r$ for fractional n by the Bernoulli-Faulhaber formula is equivalent to polynomial fitting. This choice of generalization coincides with the Euler-Maclaurin summation formula as proved in Theorem 6.2.

Exercise 6.5 Let $H(n) = \prod_{k=1}^{n} k^k$ be the *hyperfactorial* function. Use Theorem 6.2 to derive the infinite product formula:

$$\prod_{k=1}^{n} k^k = 4\, e^{n(n-1)/2 - 1} \prod_{k=2}^{\infty} \frac{(k-1)^n (1-\frac{1}{k})^{-k(n-2)-n(n-1)/2} (k+1)^{k+1}}{k(k-1)(k+n-1)^{k+n-1}} \tag{6.2.16}$$

From this equation, conclude that:

$$\prod_{k=2}^{\infty}(1+\frac{1}{k})(1-\frac{1}{k^2})^k = \frac{e}{4} \qquad (6.2.17)$$

And:

$$\sum_{k=2}^{\infty}\frac{1}{k}+k\log\left(1-\frac{1}{k^2}\right)=\gamma-\log 2 \qquad (6.2.18)$$

Exercise 6.6 Show that the infinite product formula for the superfactorial function in Eq. 6.2.10 can be, alternatively, derived from Euler's infinite product formula for the factorial function.

Exercise 6.7 Use Theorem 6.2 to prove that:

$$\sum_{k=0}^{n-1}\log k! = -\frac{\gamma}{2}n(n-1)+\log\prod_{k=0}^{\infty}\frac{k!\cdot(k+1)^n}{(k+n)!}e^{\frac{n(n-1)}{2(k+1)}} \qquad (6.2.19)$$

Use this expression and fact that [CS97]:

$$\sum_{k=0}^{-\frac{3}{2}}\log k! = \frac{\log 2}{24}-\frac{\log \pi}{4}+\frac{1}{8}-\frac{3\log A}{2}, \qquad (6.2.20)$$

where $A = 1.282427\cdots$ is the Glaisher constant, to prove the following result:

$$\lim_{s\to\infty}\left\{-\frac{s}{2}(\log\pi-1)-\frac{3}{8}\log s-\frac{1}{2}s\log s+\sum_{k=0}^{s-1}(s-k)\log\frac{2k+2}{2k+1}\right\}$$
$$=\frac{7\log 2}{24}+\frac{\log\pi}{2}+\frac{1}{8}-\frac{3\log A}{2} \qquad (6.2.21)$$

6.3 Evaluating Oscillating Simple Finite Sums

6.3.1 Main Results

In Chap. 2, we proved that simple finite sums are formally equivalent to the Euler-Maclaurin summation formula. In addition, we showed that simple finite sums of the form $\sum_{k=0}^{n-1}g(k)$ in which $g(k)$ is of a finite polynomial order are easy to work with. For instance, instead of evaluating the entire Euler-Maclaurin summation formula, we can exploit the fact that $g(k)$ has a finite polynomial order and use Eq. 2.4.8 instead. Similarly, such an advantage has showed up again in Sect. 6.2, as stated in Eq. 6.2.2.

On the other hand, if $g(k)$ is oscillating, e.g. it is of the form $g(k) = s_k z(k)$ for some non-constant, periodic sign sequence s_k, then $g(k)$ does not have a finite polynomial order. Nevertheless, if $z(k)$ does, then we can use the summation formula for oscillating sums to exploit such an advantage. In fact, this was the main motivation behind introducing that family of summation formulas, which allowed us to perform many deeds with ease, such as performing infinitesimal calculus and computing asymptotic expansions.

The objective of this section is to present the analog of Theorem 6.2 for oscillating simple finite sums. Because each function $\sum_{k=0}^{n-1} s_k z(k)$, where s_k is periodic with period p, can be decomposed into a sum of a finite number of functions of the form $\sum_{k=0}^{n-1} e^{i\theta k} z(k)$, as shown in Corollary 5.2, we will focus on the latter class of functions only. Again, the first (DC) component, given by $\left(\frac{1}{p} \sum_{k=0}^{p-1} s_k\right) \sum_{k=0}^{n-1} g(k)$ is evaluated using Theorem 6.2 but all other components in which $\theta \neq 0$ are evaluated using the following Theorem.

Theorem 6.3 *Given a simple finite sum of the form* $f(n) = \sum_{k=0}^{n-1} e^{i\theta k} g(k)$, *let* $f_G(n)$ *denote the unique natural generalization implied by the successive polynomial approximation method of Theorem 2.1. Then,* $f_G(n)$ *for all* $n \in \mathbb{C}$ *is formally given by:*

$$f_G(n) = e^{i\theta s} \sum_{r=0}^{\infty} \frac{\Omega_r(n)}{r!} g^{(r)}(s) + \sum_{k=0}^{s-1} e^{i\theta k} g(k) - e^{i\theta(k+n)} g(k+n) \qquad (6.3.1)$$

Here, $\Omega_r(n) = \sum_{k=0}^{n-1} e^{i\theta k} k^r$, *i.e. an "oscillating polynomial", whose closed-form expression is given by:*

$$\Omega_r(n) = \Theta_r - e^{i\theta n} \sum_{m=0}^{r} \binom{r}{m} \Theta_m n^{r-m}, \qquad (6.3.2)$$

where Θ_r *are the family of sequences of Theorem 5.3.*

Proof Similar to the proof of Theorem 6.2. The closed-form expression in Eq. 6.3.2 is obtained from Theorem 5.3.   □

6.3.2   Examples

As an example to Theorem 6.3, suppose we would like to evaluate the alternating simple finite sum $\sum_{k=0}^{\frac{1}{2}} (-1)^k \log(1+k)$. Because $\log(1+k)$ is of a finite polynomial order zero, we have:

$$\sum_{k=0}^{1/2}(-1)^k \log(1+k) = \lim_{s\to\infty}\left\{(-1)^s\,\Omega_0\left(\frac{1}{2}\right)\log(1+s)\right.$$

$$\left. + \sum_{k=0}^{s-1}(-1)^k \log(k+1) - (-1)^{k+\frac{3}{2}}\log\left(k+\frac{5}{2}\right)\right\}$$

(6.3.3)

However, by know by definition that $\Omega_0(x) = \sum_{k=0}^{x-1}(-1)^k = \frac{1}{2}(1-(-1)^x)$. Thus, we have:

$$\sum_{k=0}^{\frac{1}{2}}(-1)^k \log(1+k) = \lim_{s\to\infty}\left\{(-1)^s\frac{1+i}{2}\log(1+s)+\right.$$

$$\left. \sum_{k=0}^{s-1}(-1)^k \log(k+1) - (-1)^{k+\frac{3}{2}}\log\left(k+\frac{5}{2}\right)\right\}$$

Numerically, this approach yields the value $\sum_{k=1}^{\frac{1}{2}}(-1)^k \log k \approx -0.2258+i\,0.3604$. This is indeed the correct value as will be verified shortly, where the real part is given by $(\log 2 - \log \pi)/2$.

Whereas Theorem 6.3 is occasionally useful, it is often easier in practice to work with the summability identity:

$$\sum_{k=0}^{n-1}e^{i\theta k}\,g(k) = \sum_{k=0}^{\infty}e^{i\theta k}\,g(k) - \sum_{k=n}^{\infty}e^{i\theta k}\,g(k)$$

(6.3.4)

Here, all infinite sums are interpreted using the $\mathfrak{T}$ definition and can be calculated using χ. For example, suppose we would like to evaluate the fractional finite sum $\sum_{k=0}^{\frac{1}{2}}(-1)^k \log(1+k)$ addressed above. Using summability theory, we have:

$$\sum_{k=0}^{\frac{1}{2}}(-1)^k \log(1+k) = \sum_{k=0}^{\infty}(-1)^k \log(1+k) - \sum_{k=\frac{3}{2}}^{\infty}(-1)^k \log(1+k)$$

Numerically, this gives the value $-0.2258 + i\,0.3604$, which is similar to what was obtained earlier using Theorem 6.3.

Exercise 6.8 Prove that the real part of $\sum_{k=0}^{\frac{1}{2}}(-1)^k \log(1+k)$ is given by $(\log 2 - \log \pi)/2$. (Hint: use Eq. 6.3.4 to isolate the real-part and apply the results of Chap. 5)

6.4 Evaluating Composite Finite Sums

Finally, we extend the main results to the case of composite finite sums.

6.4.1 Main Results

Theorem 6.4 *Let $\sum_{k=0}^{n-1} e^{i\theta k} g(k, n)$ be a composite finite sum for some fixed $\theta \in \mathbb{R}$ and $g(\cdot, 0)$ is regular at the origin with respect to its first argument. Then:*

$$f_G(n) = e^{i\theta s} \sum_{r=0}^{\infty} \frac{\Omega_r(n)}{r!} \frac{\partial^r}{\partial s^r} g(s, n) + \sum_{k=0}^{s-1} e^{i\theta k} g(k, n) - e^{i\theta(k+n)} g(k + n, n),$$

(6.4.1)

where $\Omega_r(n) = \sum_{k=0}^{n-1} e^{i\theta k} k^r$.

Proof Similar to the proofs of the earlier theorems. However, we note here that no recurrence identity holds so the proof consists of two parts: (1) showing that the initial condition holds, and (2) showing that the derivative of the expression in Eq. 6.4.1 coincides with the results obtained in Chap. 3.   □

Remark 6.2 If $g(k, n)$ has a finite polynomial order m with respect to its first argument, then the infinite sum can be evaluated up to m only and the overall expression is taken as an asymptotic expansion with s tending to infinity. This is similar to earlier statements.

6.4.2 Examples

Theorem 6.4 generalizes the earlier results in this chapter to the broader class of composite oscillatory finite sums. For example, suppose we wish to evaluate the composite finite sum $\sum_{k=0}^{n-1} \log\left(1 + \frac{k}{n+1}\right)$. Since $g(k, n)$ has a finite polynomial order zero with respect to its first argument, we can evaluate $f_G(n)$ for all $n \in \mathbb{C}$ using the following expression:

$$f_G(n) = \lim_{s \to \infty} \left\{ n \log\left(1 + \frac{s}{n+1}\right) + \sum_{k=0}^{s-1} \left[\log\left(1 + \frac{k}{n+1}\right) - \log\left(1 + \frac{k+n}{n+1}\right) \right] \right\}$$

(6.4.2)

For example, if $n = 2$, Eq. 6.4.2 evaluates to 0.2877, which we know is correct because $f_G(2) = \log 4/3$. In general, we know in this particular example that $f_G(n) = \log(2n)! - \log n! - n \log(1 + n)$. Therefore, we can also test Eq. 6.4.2 for

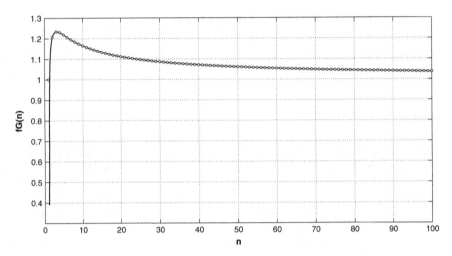

Fig. 6.2 The generalized definition of the composite finite sum $f_G(n) = \frac{1}{n}\sum_{k=1}^{n} k^{\frac{1}{n}}$ plotted against $n > 1$ using Theorem 6.4

fractional values of n. For instance, if $n = \frac{1}{2}$, then Eq. 6.4.2 evaluates to -0.0819, which is also correct because $f_G(\frac{1}{2}) = -\frac{1}{2}\log\frac{3}{2} - \log\frac{\sqrt{\pi}}{2}$.

A second example would be the composite finite sum $\frac{1}{n}\sum_{k=1}^{n} k^{\frac{1}{n}}$. Because:

$$\lim_{n\to\infty} n^{\frac{1}{n}} = 1,$$

we expect $f_G(n) \to 1$ as $n \to \infty$. In addition, $g(k) = k^{\frac{1}{n}}$ has a finite polynomial order zero if $n > 1$ so the composite sum can be evaluated using:

$$\frac{1}{n}\sum_{k=1}^{n} k^{\frac{1}{n}} = \lim_{s\to\infty} s^{\frac{1}{n}} + \frac{1}{n}\sum_{k=1}^{s+1} k^{\frac{1}{n}} - (k+n)^{\frac{1}{n}} \text{ if } n > 1 \qquad (6.4.3)$$

The function is plotted in Fig. 6.2. In the same figure, the *exact* values of $f(n)$ for $n \in \mathbb{N}$ are highlighted as well. Clearly, despite the apparent complexity of such a composite finite sum, it is indeed a simple function to compute for $n \in \mathbb{C}$. In addition, we have $f_G(n) \to 1$ when $n \to \infty$ as expected.

6.5 Summary

In this chapter, we presented a method for calculating the values of simple and composite finite sums. Whereas earlier analysis revealed that summability calculus provided an elementary framework for handling finite sums, such as performing infinitesimal calculus, deducing asymptotic expansions, as well as summing divergent series, all within a single coherent framework, the results of this chapter reveal that even the task of *calculating* composite finite sums of the general form $\sum_{k=0}^{n-1} s_k \, g(k, n)$ for $n \in \mathbb{C}$, where s_k is an arbitrary periodic sign sequence, is straightforward to implement as well. We have also proved that these methods are consistent with the earlier analysis. In the following chapter, we expand these results further using the calculus of finite differences, in which equivalent alternative results can be expressed using the language of *finite differences* as opposed to infinitesimal derivatives.

References

[CS97] J. Choi, H.M. Srivastava, Sums associated with the zeta function. J. Math. Anal. Appl. **206**(1), 103–120 (1997)

[DM09] R.J. Dwilewicz, J. Mináč, Values of the Riemann zeta function at integers. Mater. Math. **2009**(6), 1–26 (2009)

[Lam01] V. Lampret, The Euler-Maclaurin and Taylor formulas: twin, elementary derivations. Math. Mag. **74**(2), 109–122 (2001)

[MS10] M. Müller, D. Schleicher, Fractional sums and Euler-like identities. Ramanujan J. **21**(2), 123–143 (2010)

[Spo94] J.L. Spouge, Computation of the gamma, digamma, and trigamma functions. SIAM J. Numer. Anal. **31**(3), 931–944 (1994)

[Weia] E. Weisstein, Barnes G-function. http://mathworld.wolfram.com/BarnesG-Function.html

[Weib] E. Weisstein, Hyperfactorial. http://mathworld.wolfram.com/Hyperfactorial.html

Chapter 7
The Language of Finite Differences

> *What nature demands from us is not a quantum theory or a*
> *wave theory; rather, nature demands from us a synthesis of these*
> *two views.*
>
> Albert Einstein (1879–1955)

Abstract Throughout the previous chapters, we derived results using infinitesimal calculus, such as the Euler-Maclaurin summation formula, the Boole summation formula, and the methods of summability of divergent series. In this chapter, we derive analogous results using the language of finite differences, as opposed to infinitesimal derivatives. Using finite differences and the theory of summability of divergent series, we present a simple pictorial proof to the Shannon-Nyquist sampling theorem. Finally, we derive many identities that relate to the Euler constant, the Riemann zeta function, the Gregory coefficients, and the Cauchy coefficients, among others.

Throughout the previous chapters, we derived results using infinitesimal calculus, such as the Euler-Maclaurin summation formula, the Boole summation formula, and the methods of summability of divergent series. In this chapter, we derive analogous results using the language of *finite differences*, as opposed to infinitesimal derivatives.

For example, we will re-establish the well-known result that the analog of the Euler-Maclaurin summation formula for finite differences is given by the Gregory quadrature formula [Jor65, SP11]. In addition, we will show that summability calculus presents a simple, geometric proof to the well-known Shannon-Nyquist sampling theorem [Luk99]. Moreover, we will use these results to derive Newton's interpolation formula, which, as it turns out, is intimately connected to the summability method χ of Chap. 4. All of these results will be used to deduce many interesting identities related to fundamental constants, such as π, e, and the Euler constant γ.

Before we derive such results, we introduce the following notation. Throughout this chapter, we write D to denote the *differential* operator at the origin; that is,

© Springer International Publishing AG 2018
I. M. Alabdulmohsin, *Summability Calculus*,
https://doi.org/10.1007/978-3-319-74648-7_7

$D^r = f^{(r)}(0)$ are the higher order derivatives. Similarly, we write Δ to denote the *forward difference* operator at the origin; that is, $\Delta = f(1) - f(0)$, $\Delta^2 = f(2) - 2f(1) + f(0)$, and so on. When we write an expression such as $D\Delta$, it should be interpreted as the value that results from taking the derivative of the forward difference operator; i.e. $D\Delta = f'(1) - f'(0)$. Obviously, we have $D\Delta = \Delta D$. Finally, whenever we write a function that involves a symbol, such as $\log(1 + \Delta)$, it must be interpreted as a formal power series on Δ. For instance, $\log(1 + \Delta) = \sum_{k=1}^{\infty}(-1)^{k+1}\Delta/k$ whereas $e^D = \sum_{k=0}^{\infty}D^k/k!$. We always have the identity $D^0 = \Delta^0 = 1$, which is interpreted as $f(0)$, *not* the number one.

7.1 Summability Calculus and Finite Differences

Our starting point into the calculus of finite differences is to revisit the differentiation rule of simple finite sums given by Eq. 2.4.1. Using the symbols Δ and D described earlier, this is rewritten as:

$$D = \sum_{r=0}^{\infty} \frac{B_r}{r!}\Delta D^r \qquad (7.1.1)$$

Now, we note that $\Delta^p D^m$ can also be expressed using the exact same expression in Eq. 7.1.1; namely:

$$\Delta^p D^m = \sum_{r=0}^{\infty} \frac{B_r}{r!}\Delta^{p+1}D^{m+r-1} \qquad (7.1.2)$$

We also note that Eq. 7.1.2 can be used inside Eq. 7.1.1. Carrying out such a process term-by-term suggests the following formal expression for a function's derivative:

$$D = \frac{B_0}{0!}\Delta + \left(\frac{B_1}{1!}\frac{B_0}{0!}\right)\Delta^2 + \left(\frac{B_1}{1!}\frac{B_1}{1!}\frac{B_0}{0!} + \frac{B_2}{2!}\frac{B_0}{0!}\frac{B_0}{0!}\right)\Delta^3 \cdots$$

$$= \frac{\Delta}{1} - \frac{\Delta^2}{2} + \frac{\Delta^3}{3} - \frac{\Delta^4}{4} + \cdots \qquad (7.1.3)$$

Equation 7.1.3 is well-known in the calculus of finite differences (see for instance [Jor65]). We write it succinctly as:

$$D = \log(1 + \Delta) \qquad (7.1.4)$$

This seemingly simple identity is quite rich and can be used to derive many fundamental results in calculus. Because the composition of the symbols Δ and D behaves as ordinary multiplication, i.e. $D^b D^c = D^{b+c}$ and $\Delta^b \Delta^c = \Delta^{b+c}$, we deduce that for any function z:

$$z(D) = z(\log(1 + \Delta)) \tag{7.1.5}$$

Again, both sides of the above equation are interpreted as formal power series. We describe some of the implications of Eq. 7.1.5 next.

7.1.1 Higher Order Approximations to Derivatives

Given a function $f(x)$, let $g(x) = f(h x)$ for some constant $h > 0$. Applying Eq. 7.1.5 on the function g gives us the following formal expression:

$$f^{(r)}(0) = D^r = \frac{\log^r (I + \Delta_h)}{h^r} = \frac{1}{h^r}\left(\Delta_h^r - \frac{r}{2}\Delta_h^{r+1} + \cdots\right) \tag{7.1.6}$$

where $\Delta_h = f(h) - f(0)$, and $\Delta_h^r = \Delta\Delta_h^{r-1}$. So, using Eq. 7.1.6, we have a simple method for obtaining higher-order approximations to infinitesimal derivatives. One particularly important case that we will repeatedly use in the sequel is the following equation:

$$f'(x) = \frac{1}{h}\left(\frac{\Delta_h}{1} - \frac{\Delta_h^2}{2} + \frac{\Delta_h^3}{3} - \frac{\Delta_h^4}{4} + \cdots\right) \tag{7.1.7}$$

7.1.2 Taylor Expansions

If we take the composition of both sides of Eq. 7.1.6 with the exponential function $z(x) = e^x$ with $r = 1$, we have:

$$I + \Delta_h = f(x + h) = e^{hD} = I + \frac{h D}{1!} + \frac{h^2 D^2}{2!} + \frac{h^3 D^3}{3!} + \cdots \tag{7.1.8}$$

This is the Taylor series expansion of the function $f(x)$ around the origin. If, on the other hand, we select $r = 2$, we obtain:

$$e^{h^2 D^2} = (I + \Delta_h)^{\log (I + \Delta_h)} \tag{7.1.9}$$

As mentioned earlier, this is interpreted as a formal power series on both sides. That is, we have the following equation, which can be confirmed numerically quite readily.

$$I + \frac{h^2 D^2}{1!} + \frac{h^4 D^4}{2!} + \frac{h^6 D^6}{3!} + \cdots = I + \Delta_h^2 - \Delta_h^3 + \frac{17}{4}\Delta_h^4 - \frac{11}{6}\Delta_h^5 + \cdots \tag{7.1.10}$$

7.1.3   Newton's Interpolation Formula

Starting from Eq. 7.1.8, we have:

$$(I + \Delta_h)^\alpha = e^{\alpha h D} = f(\alpha h) \tag{7.1.11}$$

Equation 7.1.11 is a concise statement of Newton's interpolation formula. To see this, we select $\alpha = \frac{x}{h}$, which yields by the binomial theorem:

$$f(x) = f(x) + \frac{\Delta_h}{1!}\left(\frac{x}{h}\right) + \frac{\Delta_h^2}{2!}\left(\frac{x}{h}\right)\left(\frac{x}{h} - 1\right) + \cdots \tag{7.1.12}$$

Now, if we denote $n = \frac{x}{h}$, where h is chosen such that n is an integer, then Eq. 7.1.12 can be rewritten as:

$$f(x) = \sum_{j=0}^{n} \chi_n(j)\frac{\Delta_h^j}{h^j}\frac{x^j}{j!} \tag{7.1.13}$$

Interestingly, the function $\chi_n(j)$ used earlier in Chap. 4 shows up again. It is important to note that Eq. 7.1.13 is an *exact* expression that holds for all $n \in \mathbb{N}$, where, again, the sampling interval h is selected such that n is an integer. Letting $h \to 0$ yields Eq. 7.1.14. Note here that Newton's interpolation formula given by Eq. 7.1.14 is the discrete mirror of the summability method χ of Chap. 4, where infinitesimal derivatives $f^{(j)}(0)$ are replaced with their approximations Δ_h^j/h^j, thus bridging the two results.

$$f(x) = \lim_{n\to\infty}\left\{\sum_{j=0}^{n} \chi_n(j)\frac{\Delta_h^j}{h^j}\frac{x^j}{j!}\right\} \tag{7.1.14}$$

7.1.4   Fractional Finite Sums

Let D and Δ correspond to the function $g(k)$. We have the following formal expression:

$$-\frac{1}{\Delta} = \sum_{k=0}^{\infty} g(k), \tag{7.1.15}$$

This can be proved by applying the forward difference operator on both sides of the equation. Therefore, a simple finite sum $\sum_{k=0}^{n-1} g(k)$ can be symbolically written as:

$$\sum_{k=0}^{n-1} g(k) = \frac{(1+\Delta)^n - 1}{\Delta} = \frac{e^{nD} - 1}{e^D - 1}$$

By expanding this expression as a formal power series on D, we recover Theorem 6.2 in Chap. 6. If we write the above equation as:

$$\sum_{k=0}^{n-1} g(k) = \frac{(I + \Delta)^n - I}{e^D - 1},$$

we obtain the Euler-Maclaurin summation formula. Here, $(I + \Delta)^n$ is the shift operator, which is often denoted by E^n in the literature.

7.1.5   The Sampling Theorem

So far, we have derived higher order approximations of derivatives as well as the Newton interpolation formula, which, respectively, are the discrete analog of differentiation and computing Taylor series expansions in infinitesimal calculus. Before we start applying these results, we need to return to the expression given by Eq. 7.1.4, which was the building block of all subsequent analysis. In this equation, the well-known pitfall with *undersampling* clearly arises. For example, if $f(x) = \sin \pi x$ and $h = 1$, then $\Delta_h^j = 0$ and the right-hand side of Eq. 7.1.6 evaluates to the zero function even though the correct derivative is not. This arises because the simplest function that interpolates the samples $f(k) = 0$ is itself the zero function. Thus, our goal is to know for which sampling interval h does Eq. 7.1.6 hold? The answer is given by the following theorem.

Theorem 7.1 (The Sampling Theorem)   *Given a band-limited function $f(x)$ that is analytic over the ray $[0, \infty)$, if $f(x)$ has a bandwidth B and the sampling interval h satisfies the Nyquist criteria $h < \frac{1}{2B}$, then the $\mathfrak{T}$ definition of the series*
$\frac{1}{h} \sum_{j=0}^{\infty} (-1)^j \frac{\Delta_h^j}{j}$ *is equal to the function's derivative D. In general, Eq. 7.1.6 holds, where the right-hand series is interpreted using the generalized definition $\mathfrak{T}$.*

Proof Let the Fourier transform of the function $f(x)$ be denoted $\mathcal{F}\{f\} = \mathbf{F}(w)$. Also, let $c_{r,j}$ be the jth coefficient of the series expansion of the function $\log^r (1 + x)$. For example, we have $c_{1,j} = \frac{(-1)^{j+1}}{j}$. Then, we have by linearity of the Fourier transform[1]:

$$\mathcal{F}\left\{ \frac{1}{h^r} \sum_{j=0}^{\infty} c_{r,j} \Delta_h^j \right\} = \frac{1}{h^r} \sum_{j=0}^{\infty} c_{r,j} \mathcal{F}\{\Delta_h^j\} \qquad (7.1.16)$$

---

[1]To make this equation rigorous, one may use a summability method such as the Mittag-Leffler summability method, mentioned in Chap. 4, and use the fact that the method converges uniformly in any compact region in the interior of the Mittag-Leffler star of function.

However, if $\mathcal{F}\{f\} = \mathbf{F}(w)$, we know that $\mathcal{F}\{\Delta_h^j\} = \mathbf{F}(w)\,(e^{\mathrm{i}\,wh} - 1)^j$. Plugging this into the earlier equation yields:

$$\mathcal{F}\left\{\frac{1}{h^r}\sum_{j=0}^{\infty} c_{r,j}\,\Delta_h^j\right\} = \frac{\mathbf{F}(w)}{h^r}\sum_{j=0}^{\infty} c_{r,j}\,(e^{\mathrm{i}\,wh} - 1)^j \tag{7.1.17}$$

Because the $\mathfrak{T}$ definition correctly sums any power series in the Mittag-Leffler star of the corresponding function, we know that Eq. 7.1.18 holds as long as the line segment $[1, e^{\mathrm{i}\,wh}]$ over the complex plane $\mathbb{C}$ does not contain the origin.

$$\sum_{j=0}^{\infty} c_{r,j}\,(e^{\mathrm{i}\,wh} - 1)^j = \log^r(e^{\mathrm{i}\,wh}) = (\mathrm{i}\,wh)^r \tag{7.1.18}$$

Because $\mathbf{F}(w) = 0$ for all $|w| > B$, we need to guarantee that the line segment $[1, e^{\mathrm{i}\,wh}]$ does not pass through the origin for all $|w| \leq B$. However, to guarantee that such a condition holds, the sampling interval h has to satisfy the Nyquist criteria $h < \frac{1}{2B}$ because:

$$wh < \pi \quad \text{for } 0 \leq w \leq 2\pi B \quad \Rightarrow \quad h < \frac{1}{2B}$$

Therefore, if $f(x)$ is band-limited with bandwidth B and if $h < \frac{1}{2B}$, then the Fourier transform of the right-hand side of Eq. 7.1.6 evaluates to $(\mathrm{i}\,w)^r\,\mathbf{F}(w)$, which is also the Fourier transform of $f^{(r)}(x)$. Since the Fourier transform is invertible, both sides of Eq. 7.1.6 must be equivalent under the stated conditions when series are interpreted using the generalized definition $\mathfrak{T}$. □

Remark 7.1 Note from the proof of Theorem 7.1 that the strict inequality in the Nyquist criteria can be relaxed to an inequality $h \leq \frac{1}{2B}$ if the function $f(x)$ has no bandwidth components at frequency $w = B$.

In order to see why the generalized definition $\mathfrak{T}$ is important in our sampling theorem, we start from Eq. 7.1.5 and apply the function $z(x) = \frac{1}{x}$ to both sides, which yields:

$$\frac{1}{I + \Delta_h} = e^{-hD} \tag{7.1.19}$$

If $f(x) = e^x$ and $h = \log 2$, then the left-hand side evaluates to:

$$\frac{1}{I + \Delta_h} = I - \Delta_h + \Delta_h^2 - \Delta_h^3 + \cdots = 1 - 1 + 1 - 1 + 1 - \cdots \tag{7.1.20}$$

Of course, this series is divergent but its $\mathfrak{T}$ value is $\frac{1}{2}$ as discussed earlier. On the other hand, the right-hand side of Eq. 7.1.19 evaluates to:

$$e^{-hD} = I - hD + \frac{h^2 D^2}{2!} - \frac{h^3 D^3}{3!} + \cdots = e^{-\log 2} = \frac{1}{2} \tag{7.1.21}$$

Therefore, both sides agree as expected when interpreted using the generalized definition of infinite sums $\mathfrak{T}$. This examples demonstrates that the generalized definition of series $\mathfrak{T}$ is necessary.

Remark 7.2 Note that the sampling theorem proved in Theorem 7.1 is slightly stronger than classical versions of the theorem. In Theorem 7.1, the samples that satisfy the Nyquist criteria have to extend infinitely into the future, but *not* necessarily into the past. In other words, a function $f(x)$ can be perfectly reconstructed from its discrete samples if the discrete samples are taken in the ray $[0, \infty)$ and they satisfy the Nyquist criteria. By contrast, the classical version of the Shannon-Nyquist sampling theorem states that samples have to be taken from $-\infty$ to $+\infty$.[2]

An illustrative example is depicted in Fig. 7.1, where reconstruction is performed using Newton's interpolation formula. In this figure, the samples are taken from the function $\sin(x)$ with the sampling interval $h = \frac{1}{2}$. Clearly, not only do the 5th degree, 10th degree, and 20th degree approximations work extremely well for $x \geq 0$, but also the function's past can be reconstructed perfectly as higher degree approximations are used.

Exercise 7.1 Derive the classical Cauchy definition of a function's derivative using Eq. 7.1.6. That is, use Eq. 7.1.6 to prove that $f'(0) = \lim_{h \to 0} \{(f(h) - f(0))/h\}$.

Exercise 7.2 Use the symbolic proof techniques of this section to derive the Euler summation method of Proposition 4.6. (Hint: you may find Eq. 7.1.15 useful)

Exercise 7.3 Use Eq. 7.1.4 with $f(x) = \log(1 + x)$ to prove the following identity:

$$e = \left(\frac{2}{1}\right)^{\frac{1}{1}} \left(\frac{2^2}{1 \cdot 3}\right)^{\frac{1}{2}} \left(\frac{2^3 \cdot 4}{1 \cdot 3^3}\right)^{\frac{1}{3}} \left(\frac{2^4 \cdot 4^4}{1 \cdot 3^6 \cdot 5}\right)^{\frac{1}{4}} \cdots \tag{7.1.22}$$

---

[2]Historically, extensions of the sampling theorem have generally focused on two aspects:

1. Knowing how our a priori information about the function can reduce the required sampling rate. Examples include sampling non-bandlimited signals, which is commonly referred to as *multiresolution* or *wavelet sampling theorems* (see for instance [Vai01]).
2. Knowing how *side information* that are transmitted *along with* the discrete samples can reduce the required sampling rate. For example, whereas the bandwidth of $\sin^3(2\pi x)$ is three-times larger than the bandwidth of $\sin(2\pi x)$, we can transmit the function $\sin(2\pi x)$ along with some side information telling the receiver that the function should be cubed. Examples of this type of generalization are discussed in [Zhu92].

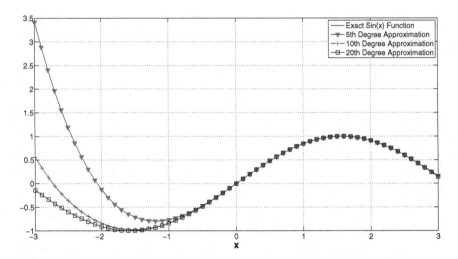

Fig. 7.1 Newton's interpolation formula applied to the discretized function $f(x) = \sin x$ with the sampling interval $h = \frac{1}{2}$ and samples are only taken in the interval $[0, \infty)$. Here, the original sine function is indistinguishable from the 20th degree approximation in the interval $-\pi \le x \le \pi$

This identity was recently proved in [GS08]. In [GM05], the same identity is both proved and generalized to $e^{\frac{1}{x}}$ for $x \in \mathbb{N}$ using probability theory. We will derive other similar identities later in Sect. 7.3.1.

Exercise 7.4 Newton's interpolation formula can be expressed concisely using the *falling factorials*. Let $(x)_r = x(x-1)(x-2)\cdots(x-r+1)$ be the falling factorial. Then, Newton's interpolation formula can be written as:

$$f(x) = \sum_{k=0}^{\infty} \frac{\Delta^k}{k!}(x)_k$$

Show that one can replace the forward difference operator with the backward difference operator ∇, where $\nabla = f(0) - f(-1)$, $\nabla^2 = f(0) - 2f(-1) + f(-2),\ldots$ etc., by replacing the falling factorials with the *rising factorials* $(x)^r = x(x+1)(x+2)\cdots(x+r-1)$. Use the fact that $\zeta(-s) = -B_{s+1}/(s+1)$ and the recurrence identity of the Bernoulli numbers to prove that:

$$\sum_{k=0}^{\infty} \frac{(-1)^{k+1}}{k^s} = \sum_{r=1}^{\infty} \frac{\zeta(r+s)\,2^{-(r+s)}}{r!}(s)^r \qquad (7.1.23)$$

This identity is attributed to Ramaswami [LOC02].

7.2 Summability Calculus Using Finite Differences

In this section, we use finite differences to establish analogs of the many summation formulas derived in earlier chapters.

7.2.1 Summation Formulas

7.2.1.1 Gregory Quadrature Formula

Our starting point is Eq. 7.1.6. If we set $r = -1$, we obtain:

$$D^{-1} = \frac{h}{\log{(I + \Delta_h)}} = h\left(\Delta_h^{-1} + \frac{1}{2}\Delta_h^0 - \frac{1}{12}\Delta_h^1 + \frac{1}{24}\Delta_h^2 + \cdots\right) \qquad (7.2.1)$$

Choosing $h = 1$ and using Eq. 7.1.15, we obtain the Gregory quadrature formula given by Eq. 7.2.2. Here, the constants $G_r = \{1, \frac{1}{2}, -\frac{1}{12}, \frac{1}{24}, -\frac{19}{720}, \frac{3}{160}, \cdots\}$ are called the Gregory coefficients (see OEIS A002206 (numerators) and A002207 (denominators) [Slod, Slob]).

$$\sum_{k=0}^{n-1} g(k) = \int_0^n g(t)\,dt + \sum_{r=1}^{\infty} G_r\left[\Delta^{r-1}g(0) - \Delta^{r-1}g(n)\right] \qquad (7.2.2)$$

Many methods exist for computing the Gregory coefficients. One convenient recursive formula is given by [Slod, Klu24]:

$$G_n = \frac{(-1)^{n+1}}{n+1} - \sum_{k=1}^{n-1}(-1)^k \frac{G_{n-k}}{k+1}, G_0 = 1 \qquad (7.2.3)$$

Clearly, Gregory's formula is analogous to the Euler-Maclaurin summation formula, where finite differences replace infinitesimal derivatives.

7.2.1.2 Oscillating Simple Finite Sums

To deduce the analog of the summation formulas for oscillating simple finite sums, we note again that the discrete analog of Taylor's series expansions is given by Newton's interpolation formula:

$$f(x) = \frac{\Delta^0}{0!} + \frac{\Delta}{1!}(x)_1 + \frac{\Delta^2}{2!}(x)_2 + \frac{\Delta^3}{3!}(x)_3 + \cdots \qquad (7.2.4)$$

Here, $(z)_n$ is the *falling factorial* defined by $(z)_n = z(z-1)(z-2)\cdots(z-n+1)$. One way of interpreting the previous equation is to note that it yields consistent results when the forward difference operator Δ is applied to both sides of the equation, which, in turn, follows because $\Delta(z)_n = n\,(z)_{n-1}$.

Using analytic summability theory, and in a manner that is analogous to the development of Chap. 5, we have:

$$\sum_{k=0}^{\infty} e^{i\theta k} g(k) = \sum_{r=0}^{\infty} \frac{\Phi_r}{r!} \Delta^r, \quad \text{where } \Phi_r = \sum_{k=0}^{\infty} e^{i\theta k}(k)_r \qquad (7.2.5)$$

Here, we note that Φ_r is the $\mathfrak{T}$ value of a divergent series that involves the falling factorials $(k)_r$. Because $(k)_r$ is a polynomial of degree r, the constants Φ_r can be computed from the constants $\Theta_r = \sum_{k=0}^{\infty} e^{i\theta k} k^r$. In particular, using the Stirling numbers of the first kind $s(r, k)$, we have:

$$\Phi_r = \sum_{k=0}^{r} s(r, k)\,\Theta_k \qquad (7.2.6)$$

Equation 7.2.5, in turn, implies that:

$$\sum_{k=0}^{n-1} e^{i\theta k} g(k) = \sum_{r=0}^{\infty} \frac{\Phi_r}{r!} \left(\Delta^r g(0) - e^{i\theta n}\,\Delta^r g(n) \right) \qquad (7.2.7)$$

In addition, we have that the $\mathfrak{T}$ value of the series $\sum_{k=0}^{\infty} e^{i\theta k} g(k)$ is formally given by:

$$\sum_{k=0}^{\infty} e^{i\theta k} g(k) = \sum_{k=0}^{n-1} e^{i\theta k} g(k) + \sum_{r=0}^{\infty} \frac{\Phi_r}{r!} e^{i\theta n}\,\Delta^r g(n) \qquad (7.2.8)$$

Equally important, the last equation is another method of deriving asymptotic expansions to oscillating sums $\sum_{k=0}^{n-1} e^{i\theta k} g(k)$ using finite differences.

7.2.2   Computing Finite Sums

Finally, to evaluate simple finite sums directly in a manner that is analogous to the development of Chap. 6, we can use the following equation, which holds formally for all s:

$$\sum_{k=0}^{n-1} g(k) = \sum_{r=0}^{\infty} \frac{d_r(n)}{r!} \Delta^r g(s) + \sum_{k=0}^{s-1} g(k) - g(k+n) \qquad (7.2.9)$$

Here, $d_r(n) = \sum_{k=0}^{n-1}(k)_r$. The proof of Eq. 7.2.9 is analogous to the proof of Theorem 6.2, and we leave it as an exercise to the reader. If $\Delta^{m+1}g(s) \to 0$ as $s \to \infty$, then Eq. 7.2.9 can be used as an asymptotic expression.

Similarly, we have the following evaluation method for oscillating finite sums:

$$\sum_{k=0}^{n-1} e^{i\theta k}g(k) = e^{i\theta s}\sum_{r=0}^{\infty} \frac{\Upsilon_r(n)}{r!}\Delta^r g(s) + \sum_{k=0}^{s-1} e^{i\theta k}g(k) - e^{i\theta(k+n)}g(k+n)$$

(7.2.10)

Here, $\Upsilon_r(x) = \sum_{k=0}^{x-1} e^{i\theta k}(k)_r$, which is available in analytic closed form. In particular, if $\Omega_r(x)$ is defined as in Theorem 6.3, then we have:

$$\Upsilon_r(x) = \sum_{k=0}^{r} s(r,k)\,\Omega_r(x)$$

(7.2.11)

Validity of the new formulas above is illustrated in the following section.

Exercise 7.5 Use the Gregory quadrature formula in Eq. 7.2.2 to derive the following series representation for the Riemann zeta function:

$$\zeta(s) = \frac{1}{s-1} - \sum_{r=1}^{\infty} G_r \sum_{k=0}^{r-1}(-1)^{r+k}\binom{r-1}{k}\frac{1}{(k+1)^s}$$

This is a globally convergent expression [Bla16b]. (Hint: start with the simple finite sum $\sum_{k=0}^{n-1} 1/k^{s-1}$, apply the Gregory quadrature formula, and take the derivatives of both sides at $n = 0$).

7.3 Applications to Summability Calculus

The rules derived in the previous section can be applied to arbitrary discrete functions. In the context of summability calculus, these rules present alternative formal methods of computing fractional finite sums. For example, we can safely replace the Euler-Maclaurin summation formula with Gregory's formula since both are formally equivalent. In addition, we can use the methods given in Eqs. 7.2.9 and 7.2.10 to compute fractional sums directly, instead of using the theorems of Chap. 6. However, we need to exercise caution when using any discretization of continuous functions to avoid undersampling!

In this section, we will present a few examples that illustrate the value of finite differences for summability calculus.

7.3.1 Binomial Coefficients and Infinite Products

In Eq. 7.1.22, we derived an infinite product formula for the natural logarithmic base e by applying Eq. 7.1.4 to the logarithmic function. Of course, the logarithmic function is merely one possibility. For instance, if we apply Eq. 7.1.7 to the simple finite sum $\sum_{k=0}^{n-1} \log(k+1)$, we obtain Eq. 7.3.1. Equation 7.3.1 was first proved by Ser in 1926 and was later rediscovered by Sondow in 2003 using hypergeometric series [Son03, GS08].

$$e^{\gamma} = \left(\frac{2}{1}\right)^{\frac{1}{2}} \left(\frac{2^2}{1 \cdot 3}\right)^{\frac{1}{3}} \left(\frac{2^3 \cdot 4}{1 \cdot 3^3}\right)^{\frac{1}{4}} \left(\frac{2^4 \cdot 4^4}{1 \cdot 3^6 \cdot 5}\right)^{\frac{1}{5}} \cdots \qquad (7.3.1)$$

Moreover, we can use the same approach to derive additional interesting identities. For example, if we apply Eq. 7.1.7 to the log-superfactorial function, $\sum_{k=1}^{n} \log k!$, we obtain:

$$\frac{e^{\gamma + \frac{1}{2}}}{\sqrt{2\pi}} = (2)^{\frac{1}{2}} \left(\frac{2}{3}\right)^{\frac{1}{3}} \left(\frac{2 \cdot 4}{3^2}\right)^{\frac{1}{4}} \left(\frac{2 \cdot 4^3}{3^3 \cdot 5}\right)^{\frac{1}{5}} \cdots \qquad (7.3.2)$$

Again, the internal exponents are given by the binomial coefficients alternating in sign. Dividing Eq. 7.3.1 by Eq. 7.3.2 yields Eq. 7.3.3. This equation was derived in [GS08] using double integrals via the analytic continuation of Lerch's Transcendent.

$$\sqrt{\frac{2\pi}{e}} = \left(\frac{2}{1}\right)^{\frac{1}{3}} \left(\frac{2^2}{1 \cdot 3}\right)^{\frac{1}{4}} \left(\frac{2^3 \cdot 4}{1 \cdot 3^3}\right)^{\frac{1}{5}} \left(\frac{2^4 \cdot 4^4}{1 \cdot 3^6 \cdot 5}\right)^{\frac{1}{6}} \cdots \qquad (7.3.3)$$

Finally, suppose we have $f(n) = \sum_{k=1}^{n} \log \frac{4k-1}{4k-3}$, and we wish to find its derivative at $n = 0$. Using summability calculus, we immediately have by Theorem 2.2 that the following equation holds:

$$f'(0) = 4 \sum_{k=1}^{\infty} \left(\frac{1}{4k-3} - \frac{1}{4k-1}\right) = 4\left(1 - \frac{1}{3} + \frac{1}{5} - \frac{1}{7} + \cdots\right) = \pi \qquad (7.3.4)$$

Using Eq. 7.1.7, on the other hand, yields the infinite product formula:

$$f'(0) = \log\left[\left(\frac{3}{1}\right)^{\frac{1}{1}} \left(\frac{3 \cdot 5}{1 \cdot 7}\right)^{\frac{1}{2}} \left(\frac{3 \cdot 5^2 \cdot 11}{1 \cdot 7^2 \cdot 9}\right)^{\frac{1}{3}} \cdots\right] \qquad (7.3.5)$$

Equating both equations yields Eq. 7.3.6. An alternative proof to this identity is given in [GS08].

$$e^{\pi} = \left(\frac{3}{1}\right)^{\frac{1}{1}} \left(\frac{3 \cdot 5}{1 \cdot 7}\right)^{\frac{1}{2}} \left(\frac{3 \cdot 5^2 \cdot 11}{1 \cdot 7^2 \cdot 9}\right)^{\frac{1}{3}} \cdots \qquad (7.3.6)$$

7.3.2 The Zeta Function Revisited

Suppose we start with the harmonic numbers $f(n) = \sum_{k=0}^{n-1} \frac{1}{1+k}$ and we wish to find its derivative at $n = 0$. We know by summability calculus that the derivative is given by $\zeta(2)$. Since we have for the harmonic numbers $\Delta^p = \frac{(-1)^{p+1}}{p}$, applying Eq. 7.1.7 also yields the original series representation for $\zeta(2)$. However, such an approach can also be applied to the generalized harmonic numbers $\sum_{k=0}^{n-1} \frac{1}{(1+k)^s}$, which would yield a different series representation for the Riemann zeta function $\zeta(s)$ as given by Eq. 7.3.7. Equation 7.3.7 follows immediately by summability calculus and Eq. 7.1.7. Interestingly, this identity is a globally convergent expression to the Riemann zeta function for all $s \in \mathbb{C}$. Variants of this formula were discovered by Helmut Hasse in 1930 and by Joseph Ser in 1926 [SW, Bla16a].

$$\zeta(s) = \frac{1}{1-s} \sum_{k=1}^{\infty} \frac{1}{k} \sum_{j=1}^{k} \binom{k-1}{j-1} \frac{(-1)^j}{j^{s+1}} \qquad (7.3.7)$$

We can also derive an integral representation of the Riemann zeta function by applying Eq. 7.1.6 to the harmonic numbers $f(n) = H_n$. First, we have $f^{(r)}(0) = (-1)^{r+1} r! \zeta(r+1)$. Second, we note by Eq. 7.1.6 and the fact that the forward differences $\Delta^{(r)}$ for $r \geq 1$ are given by the sequence $(1, -\frac{1}{2}, \frac{1}{3}, -\frac{1}{4}, \ldots)$:

$$(-1)^{r+1} r! \zeta(r+1) = \int_0^1 \frac{\log(1-x)^r}{x} dx$$

Therefore:

$$\zeta(s) = \frac{(-1)^{s+1}}{s!} \int_0^1 \frac{\log(1-x)^{s-1}}{x} dx, \qquad \Re(s) > 1 \qquad (7.3.8)$$

7.3.3 Identities Involving Gregory's Coefficients

In this example, we will use Gregory's formula to derive identities that relate the Gregory coefficients with many fundamental constants. Our starting point will be the case of *semi-linear* simple finite sums $\sum_{k=0}^{n-1} g(k)$ that have well-known asymptotic expansions $S_g(n)$ such that $\lim_{n \to \infty} \{S_g(n) - \sum_{k=0}^{n-1} g(k)\} = 0$. To recall earlier discussions, if a simple finite sum is semi-linear, then $g'(n) \to 0$ as $n \to \infty$, which implies that $\Delta g(n) \to 0$ as $n \to \infty$.

Now, by Gregory's formula:

$$\lim_{n \to \infty} \left\{ \sum_{k=0}^{n-1} g(k) + \frac{g(n)}{2} - \int_0^n g(t) \, dt \right\} = \sum_{r=1}^{\infty} G_r \Delta^{r-1} \qquad (7.3.9)$$

For example, if $g(k) = \frac{1}{k}$, then $\Delta^p = \frac{(-1)^p}{p+1}$. Plugging this into the above formula yields:

$$\gamma = \sum_{r=1}^{\infty} \frac{|G_r|}{r} = \frac{1}{2} + \frac{1}{24} + \frac{1}{72} + \frac{19}{2880} + \frac{3}{800} + \frac{863}{362880} + \cdots \qquad (7.3.10)$$

Equation 7.3.10 is one of the earliest expressions discovered that express Euler's constant γ as a limit of rational terms. It was discovered by Gregorio Fontana and Lorenzo Mascheroni in 1790 and, much later, by Kluyver in 1924 using the integral representation of the digamma function $\psi(n)$[Bla16a, Klu24].

Similarly, if we now let $g(k) = H_k$, then we have:

$$\lim_{n\to\infty} \left\{ \sum_{k=0}^{n-1} H_k + \frac{H_n}{2} - \int_0^n H_t \, dt \right\} = \lim_{n\to\infty} \left\{ (n+1)(H_{n+1} - 1) - \frac{H_n}{2} - \gamma n - \log n! \right\}$$

$$= \lim_{n\to\infty} \left\{ n\left(H_n - \log n - \gamma\right) + \frac{H_n - \log n - \log 2\pi}{2} \right\}$$

$$= \frac{1 + \gamma - \log 2\pi}{2}$$

Because $\Delta^p H_0 = \frac{(-1)^{p+1}}{p}$, we have:

$$\sum_{r=2}^{\infty} \frac{|G_r|}{r-1} = \frac{\log 2\pi - 1 - \gamma}{2} = \frac{1}{12} + \frac{1}{48} + \frac{19}{2160} + \cdots \qquad (7.3.11)$$

In addition, we have by Gregory's formula:

$$\sum_{k=0}^{-1} g(k+a) = \sum_{k=a}^{a-1} g(k) = \int_a^{a-1} g(t) \, dt + \frac{g(a) + g(a-1)}{2} + \sum_{r=2}^{\infty} G_r \Delta^r g(a-1) \qquad (7.3.12)$$

Here, we have used the fact that $\Delta^{r-1} g(a) - \Delta^{r-1} g(a-1) = \Delta^r g(a-1)$. However, by the empty sum rule, Eq. 7.3.12 implies that:

$$\int_{a-1}^{a} g(t) \, dt = \frac{g(a) + g(a-1)}{2} + \sum_{r=2}^{\infty} G_r \Delta^r g(a-1) \qquad (7.3.13)$$

The last equation allows us to deduce a rich set of identities. For example, if $g(k) = \frac{1}{k}$ and $a = 2$, we have the following identity (compare it with Eqs. 7.3.10

and 7.3.11):

$$\sum_{r=1}^{\infty} \frac{|G_r|}{r+1} = 1 - \log 2 \tag{7.3.14}$$

Similarly, using $g(k) = \sin \frac{\pi}{3}k$ or $g(k) = \cos \frac{\pi}{3}k$ yield identities such as:

$$|G_1| + |G_2| - |G_4| - |G_5| + |G_7| + |G_8| - |G_{10}| - |G_{11}| + \cdots = \frac{\sqrt{3}}{\pi} \tag{7.3.15}$$

$$|G_2| + |G_3| - |G_5| - |G_6| + |G_8| + |G_9| - |G_{11}| - |G_{12}| + \cdots = \frac{2\sqrt{3}}{\pi} - 1 \tag{7.3.16}$$

$$|G_1| - |G_3| - |G_4| + |G_6| + |G_7| - |G_9| - |G_{10}| + |G_{12}| + \cdots = 1 - \frac{\sqrt{3}}{\pi} \tag{7.3.17}$$

7.3.4   Identities Involving the Cauchy Numbers

Starting, again, from Eq. 7.1.4, we have the symbolic expression:

$$\frac{\Delta}{(I + \Delta)D} = \frac{\Delta}{(1 + \Delta) \log(1 + \Delta)} \tag{7.3.18}$$

This is interpreted by the following asymptotic expansion:

$$\int_0^n g(t-1)dt = \sum_{k=0}^{n-1} g(k) - \sum_{r=1}^{\infty} \frac{C_r}{r!}(\Delta_n^{r-1} - \Delta_0^{r-1}), \tag{7.3.19}$$

where the constants C_r have the exponential generating function $\frac{x}{(1+x)\log(1+x)} = \sum_{r=0}^{\infty} \frac{C_r}{r!} x^r$. These constants are, sometimes, referred to as the Cauchy numbers of the second kind or the Nörlund numbers (see OEIS sequence A002790 and A002657) [Sloa, Sloc].

If we apply the summation formula in Eq. 7.3.19 to the harmonic numbers, we have:

$$\gamma = \int_0^1 H_t dt = 1 - \sum_{r=0}^{\infty} \frac{(-1)^r C_r}{r(r+1)!}, \tag{7.3.20}$$

where γ is the Euler constant. This formula was recently derived by Blagouchine [Bla16b].

Applying the same formula, again, on the functions $\sin(\frac{\pi k}{3})$ and $\cos(\frac{\pi k}{3})$, we deduce the following identities:

$$\frac{|C_1|}{1!} + \frac{|C_2|}{2!} - \frac{|C_4|}{4!} - \frac{|C_5|}{5!} + \frac{|C_7|}{7!} + \frac{|C_8|}{8!} - \frac{|C_{10}|}{10!} - \frac{|C_{11}|}{11!} + \cdots = \frac{\sqrt{3}}{\pi}$$

$$\tag{7.3.21}$$

$$\frac{|C_2|}{2!} + \frac{|C_3|}{3!} - \frac{|C_5|}{5!} - \frac{|C_6|}{6!} + \frac{|C_8|}{8!} + \frac{|C_9|}{9!} - \frac{|C_{11}|}{11!} - \frac{|C_{12}|}{12!} + \cdots = 1 - \frac{\sqrt{3}}{\pi}$$

$$\tag{7.3.22}$$

$$\frac{|C_1|}{1!} - \frac{|C_3|}{3!} - \frac{|C_4|}{4!} + \frac{|C_6|}{6!} + \frac{|C_7|}{7!} - \frac{|C_9|}{9!} - \frac{|C_{10}|}{10!} + \frac{|C_{12}|}{12!} + \cdots = \frac{2\sqrt{3}}{\pi} - 1,$$

$$\tag{7.3.23}$$

which are quite similar to the identities derived earlier for the Gregory coefficients.

Exercise 7.6 Use Eq. 7.3.7 and the infinite product representation for e^γ in Eq. 7.3.1 to prove that:

$$\lim_{s \to 1} \left\{ \zeta(s) - \frac{1}{s-1} \right\} = \gamma \tag{7.3.24}$$

Exercise 7.7 Use Eq. 7.1.4 to prove that if $f(n) = H_n$ is the harmonic numbers, then $f'(0) = \zeta(2) = \pi^2/6$. (Hint: if $g(x) = 1/(1+x)$, then $\Delta^{(r)}g(0) = (-1)^r/(r+1)$).

7.4 Summary

In this chapter, summability calculus was used to derive foundational results in the calculus of finite differences, such as the higher order approximations of derivatives, the Newton interpolation formula, and the Gregory quadrature formula. Such results were, in turn, used in conjunction with analytic summability theory to derive the celebrated Shannon-Nyquist sampling theorem and to deduce the discrete analog of the many summation formulas derived in previous chapters. Also, new methods for computing fractional finite sums were established. Finally, all of these results were used to deduce many interesting identities related to fundamental constants, such as π, e, and γ.

References

[Bla16a] I.V. Blagouchine, Expansions of generalized euler's constants into the series of polynomials in and into the formal enveloping series with rational coefficients only. J. Number Theory **158**, 365–396 (2016)

[Bla16b] I.V. Blagouchine, Three notes on Ser's and Hasse's representations for the zeta-functions (2016). arXiv preprint arXiv:1606.02044

[GM05] C. Goldschmidt, J.B. Martin, Random recursive trees and the Bolthausen-Sznitman coalescent. Electron. J. Probab. **10**, 718–745 (2005). Also available at: http://arxiv.org/abs/math.PR/0502263

[GS08] J. Guillera, J. Sondow, Double integrals and infinite products for some classical constants via analytic continuations of Lerch's transcendent. Ramanujan J. **16**(3), 247–270 (2008)

[Jor65] K. Jordán, *Calculus of Finite Differences* (Chelsea Publishing Company, London, 1965)

[Klu24] J.C. Kluyver, Euler's constant and natural numbers. Proc. K. Ned. Akad. Wet **27**, 142–144 (1924)

[LOC02] H. Lee, B.M. Ok, J. Choi, Notes on some identities involving the Riemann zeta function. Commun. Korean Math. Soc. **17**(1), 165–174 (2002)

[Luk99] H.D. Luke, The origins of the sampling theorem. IEEE Commun. Mag. **37**(4), 106–108 (1999)

[SP11] S. Sinha, S. Pradhan, *Numerical Analysis & Statistical Methods* (Academic Publishers, Kolkata, 2011)

[Sloa] N. Sloane, Denominators of cauchy numbers of second type. http://oeis.org/A002790

[Slob] N. Sloane, Denominators of logarithmic numbers (also of Gregory coefficients $g(n)$). http://oeis.org/A002207

[Sloc] N. Sloane, Numerators of cauchy numbers of second type. http://oeis.org/A002657

[Slod] N. Sloane, Numerators of logarithmic numbers (also of Gregory coefficients $g(n)$). http://oeis.org/A002206

[Son03] J. Sondow, An infinite product for e^λ via hypergeometric formulas for Euler's constant (2003). arXiv:math/0306008v1 [math.CA]

[SW] J. Sondow, E. Weisstein, Riemann zeta function. http://mathworld.wolfram.com/RiemannZetaFunction.html

[Vai01] P. Vaidyanathan, Generalizations of the sampling theorem: seven decades after Nyquist. IEEE Trans. Circuits Syst. I, Fundam. Theory Appl. **48**(9), 1094–1109 (2001)

[Zhu92] Y. Zhu, Generalized sampling theorem. IEEE Trans. Circuits Syst. II, Analog Digit. Signal Process. **39**(8), 587–588 (1992)

Appendix A
The Sum of the Approximation Errors
of Harmonic Numbers

A.1 Introduction

In this appendix, we compile a list of identities that involve the sum of approximation errors of harmonic numbers by the natural logarithmic function.

Specifically, a famous result, due to Euler, is that the limit:

$$\lim_{n \to \infty} \left\{ \sum_{k=1}^{n} \frac{1}{k} - \log n \right\}$$

exists and is given by a constant $\gamma = 0.55721$. This was proved in Chap. 2. Consequently, we know that the following alternating series:

$$\sum_{k=1}^{\infty} (-1)^k \left[H_k - \log k - \gamma \right] \tag{A.1.1}$$

converges to some finite value. Our first goal is to derive an exact expression to this value.

Using the Euler-Maclaurin summation formula, it can be shown that the approximation error $H_n - \log n - \gamma$ has the asymptotic expansion:

$$H_n - \log n - \gamma \sim \frac{1}{2n} - \sum_{k=2}^{\infty} \frac{B_k}{k\,n^k}$$

This shows that the sum of approximation errors $\sum_{k=1}^{\infty} [H_k - \log k - \gamma]$ diverges. Adding an additional term from the asymptotic expansion above gives us:

$$H_n - \log n - \gamma - \frac{1}{2n} \sim O(1/n^2)$$

© Springer International Publishing AG 2018
I. M. Alabdulmohsin, *Summability Calculus*,
https://doi.org/10.1007/978-3-319-74648-7

Hence, the following series:

$$\sum_{k=1}^{\infty}\left[H_k - \log k - \gamma - \frac{1}{2k}\right] \tag{A.1.2}$$

converges to some finite value. Our second goal is to derive a closed-form expression for this series.

Finally, using the Euler-Maclaurin summation formula, Cesaro and Ramanujan, among others, deduced the following asymptotic expansion [Seb02]:

$$H_n - \log \sqrt{n(n+1)} - \gamma \sim \frac{1}{6n(n+1)}$$

As a result, the natural logarithms of the geometric means offer a more accurate approximation to the harmonic numbers. In particular, the following series:

$$\sum_{k=1}^{\infty}\left[H_k - \log \sqrt{k(k+1)} - \gamma\right] \tag{A.1.3}$$

converges to some finite value. Our third goal is to derive a closed-form expression for this series.

A.2 Main Theorem

Remarkably, for all the three series in Eqs. A.1.1–A.1.3, their exact values are expressed in terms of γ itself as well as $\log \pi$. More specifically, we have the following theorem:

Theorem A.1 *The sum of approximation errors of harmonic numbers by the natural logarithmic function satisfy the following identities:*

$$\sum_{k=1}^{\infty}\left[\log(k) + \gamma - H_k + \frac{1}{2k}\right] = \frac{\log(2\pi) - 1 - \gamma}{2}$$

$$\sum_{k=1}^{\infty}\left[\log \sqrt{k(k+1)} + \gamma - H_k\right] = \frac{\log(2\pi) - 1}{2} - \gamma$$

$$\sum_{k=1}^{\infty}(-1)^k\left[\log k + \gamma - H_k\right] = \frac{\log \pi - \gamma}{2}$$

Proof The third identity was proved in Eq. 5.1.33 using the theory of summability of divergent series. The second equation was proved in Eq. 6.2.12 using the infinite product representation of the superfactorial function. The first identity follows from the second identity because:

$$\sum_{k=1}^{\infty} \log k + \gamma - H_k + \frac{1}{2k} = \sum_{k=1}^{\infty} \log \sqrt{k(k+1)} + \gamma - H_k + \frac{1}{2}\left(\frac{1}{k} - \log\left(1 + \frac{1}{k}\right)\right)$$

$$= \frac{\log(2\pi) - 1}{2} - \gamma + \frac{\gamma}{2}$$

$$= \frac{\log(2\pi) - 1 - \gamma}{2}$$

Reference

[Seb02] P. Sebha, Collection of formulae for Euler's constant (2002). http://scipp.ucsc.edu/~haber/archives/physics116A06/euler2.ps. Retrieved on March 2011

Glossary

Abel Summability Method Given a series $\sum_{k=0}^{\infty} \alpha_k$, the Abel summability method assigns the following value if the limit exists:

$$\sum_{k=0}^{\infty} \alpha_k \triangleq \lim_{x \to 1^-} \sum_{k=0}^{\infty} \alpha_k x^k$$

The Abel summability method is consistent with, but is more powerful than, all Nörlund Means.

Bernoulli Numbers Bernoulli numbers is a sequence of rational numbers that arise frequently in number theory. Their first few terms are given by:

$$B_0 = 1, \quad B_1 = -\frac{1}{2}, \quad B_2 = \frac{1}{6}, \quad B_3 = 0, \quad B_4 = -\frac{1}{30}, \quad B_5 = 0, \ldots$$

Throughout this monograph, we adopt the convention $B_1 = -\frac{1}{2}$. Bernoulli numbers vanish at positive odd integers exceeding 1. The generating function of Bernoulli numbers is given by:

$$\frac{x}{e^x - 1} = \sum_{r=0}^{\infty} \frac{B_r}{r!} x^r$$

Composite Finite Sums A function $f(n)$ is a composite finite sum if it is of the form:

$$f(n) = \sum_{k=0}^{n-1} g(k, n)$$

I. M. Alabdulmohsin, *Summability Calculus*,
https://doi.org/10.1007/978-3-319-74648-7

Here, the iterated function $g(k, n)$ can depend on the bound n. The functions $\sum_{k=0}^{n-1} \log (1 + k/n)$ and $\sum_{k=0}^{n-1} (k + n)^{-1}$ are examples to composite finite sums.

Glaisher Approximation The Glaisher approximation is an asymptotic expression to the hyperfactorial function. It states that:

$$H(n) \sim A\, e^{-n^2/4}\, n^{n(n+1)/2+1/12},$$

with a ratio that goes to unity as $n \to \infty$. Here, $A \approx 1.2824$ is often referred to as the Glaisher-Kinkelin constant.

Hyperfactorial Function The hyperfactorial function $H(n)$ is defined for $n \in \mathbb{N}$ by the formula:

$$H(n) = \prod_{k=1}^{n} k^k$$

Lindelöf Summability Method The Lindelöf summability method assigns to a series $\sum_{k=0}^{\infty} \alpha_k$ the value:

$$\sum_{k=0}^{\infty} \alpha_k \triangleq \lim_{\delta \to 0} \sum_{k=0}^{\infty} k^{-\delta k} \alpha_k$$

It can correctly sum any Taylor series expansion in the Mittag-Leffler star.

Nearly Convergent Functions A function $f(x)$ is called nearly convergent if $\lim_{x \to \infty} f'(x) = 0$ and one of the following two conditions holds:

1. $f(x)$ is asymptotically non-decreasing and concave. More precisely, there exists x_0 such that for all $x > x_0$, we have $f'(x) \geq 0$ and $f^{(2)}(x) \leq 0$.
2. $f(x)$ is asymptotically non-increasing and convex. More precisely, there exists x_0 such that for all $x > x_0$, we have $f'(x) \leq 0$ and $f^{(2)}(x) \geq 0$

Examples of nearly convergent functions include $\sqrt{x}$, $1/x$, and $\log x$.

Polynomial Order For any function $f(x)$, its polynomial order is the *minimum* $r \geq 0$ such that:

$$\lim_{x \to \infty} \frac{d^{r+1}}{dx^{r+1}} f(x) = 0$$

For example, the function $f(x) = x\sqrt{x}$ has a polynomial order of 1 because the first derivative $\frac{3\sqrt{x}}{2}$ does not vanish as $x \to \infty$ whereas its second derivative $\frac{3}{4\sqrt{x}}$ does.

Riemann Zeta Function The Riemann zeta function $\zeta(s)$ is defined for $\mathcal{R}(s) > 1$ by the series:

$$\zeta(s) = \sum_{k=1}^{\infty} \frac{1}{k^s}$$

For positive even integers, Euler showed that the following equation holds:

$$\zeta(2s) = (-1)^{s+1} \frac{B_{2s} (2\pi)^{2s}}{2 (2s)!}$$

Semi-linear Simple Finite Sums A function $f(n)$ is a semi-linear simple finite sum if it is a simple finite sum of the form $\sum_{k=0}^{n-1} g(k)$ and $g(k)$ is a nearly-convergent function.

Simple Finite Sums A function $f(n)$ is a simple finite sum if it is of the form:

$$f(n) = \sum_{k=0}^{n-1} g(k)$$

Here, the iterated function $g(k)$ is independent of n. For example, the log-factorial function $\sum_{k=0}^{n-1} \log(1+k)$ and the harmonic number $\sum_{k=0}^{n-1} \frac{1}{1+k}$ are simple finite sums.

Stirling Approximation The Stirling approximation is an asymptotic expression for the factorial function. It states that:

$$n! \sim \sqrt{2\pi n} \left(\frac{n}{e}\right)^n,$$

with a ratio that goes to unity as $n \to \infty$.

Summability Methods A summability method is a method of defining divergent series. One simple summability method is to define a series by the following expression:

$$\sum_{k=0}^{\infty} \alpha_k \triangleq \lim_{n \to \infty} \frac{1}{2} \left(\sum_{k=0}^{n} \alpha_k + \sum_{k=0}^{n+1} \alpha_k \right)$$

This definition assigns the value $\frac{1}{2}$ to the Grandi series $1 - 1 + 1 - 1 + 1 - \ldots$, which agrees with other methods. Two summability methods are called *consistent* if they always assign the same value to a series whenever the series is well-defined under both methods. A summability method $\mathbb{T}_1$ is *stronger* than a summability method $\mathbb{T}_2$ if both are consistent and $\mathbb{T}_1$ can sum more series.

Three desirable properties of summability methods are:

1. *Regularity*: A summability method is called regular if it agrees with ordinary summation whenever a series converges in the classical sense.
2. *Linearity*: A summability method is linear if we always have:

$$\sum_i^{\infty} \alpha_i + \lambda \beta_i = \sum_i^{\infty} \alpha_i + \lambda \sum_i^{\infty} \beta_i$$

3. *Stability*: A summability method is called stable if we always have:

$$\sum_{k=0}^{\infty} \alpha_k = \alpha_0 + \sum_{k=0}^{\infty} \alpha_{1+k}$$

Superfactorial Function The superfactorial function $S(n)$ is defined for $n \in \mathbb{N}$ by the formula:

$$S(n) = \prod_{k=1}^{n} k!$$

Telescoping Sums and Products A simple finite sum of the form $\sum_{k=0}^{n-1} g(k)$ is called *telescoping* if it can be written in the form:

$$f(n) = \sum_{k=a}^{n} h(k) - h(k+1)$$

An often-cited example is the finite sum:

$$f(n) = \sum_{k=1}^{n} \frac{1}{n(n+1)}$$

This finite sum can be rewritten as:

$$f(n) = \sum_{k=1}^{n} \frac{1}{k(k+1)} = \sum_{k=1}^{n} \frac{1}{k} - \frac{1}{k+1}$$

$$= 1 - [\frac{1}{2} - \frac{1}{2}] - [\frac{1}{3} - \frac{1}{3}] - \ldots - [\frac{1}{n} - \frac{1}{n}] - \frac{1}{n+1}$$

$$= 1 - \frac{1}{n+1}$$

Hence, the sum is telescoping. A finite product $f(n) = \prod_{k=0}^{n-1} g(k)$ is called telescoping if $\log f(n)$ is a telescoping sum.

The Bohr-Mollerup Theorem The Bohr-Mollerup theorem is one possible characterization for the *uniqueness* of the gamma function in generalizing the definition of the discrete factorial function. The theorem is due to Harald Bohr and Johannes Mollerup who proved that $\Gamma(n + 1)$ is the only function that satisfies:

1. $f(n) = nf(n)$
2. $f(1) = 1$
3. $f(n)$ is logarithmically convex.

The Cesàro Means The Cesàro mean is a summability method introduced by Ernesto Cesàro. It is an averaging technique, which defines a series $\sum_{k=0}^{\infty} \alpha_k$ by:

$$\sum_{k=0}^{\infty} \alpha_k \triangleq \lim_{n \to \infty} \left\{ \frac{1}{n} \sum_{j=1}^{n} \sum_{k=0}^{j} \alpha_k \right\}$$

For convergent sums, $\sum_{k=0}^{j} \alpha_k$ tends to a limit V as $j \to \infty$ so the overall Cesàro mean tends to the same limit. Therefore, the Cesàro mean is a regular summability method. It is also linear and stable. The Cesàro mean is occasionally referred to as the Cesàro summability method.

The Euler Sum Given a series $\sum_{k=0}^{\infty} (-1)^k \alpha_k$, its Euler sum is defined by the limit:

$$\sum_{k=0}^{\infty} (-1)^k \alpha_k \triangleq \lim_{n \to \infty} \sum_{k=0}^{n} \frac{(-1)^k}{2^{k+1}} \Delta^k \alpha,$$

if the limit exists. The Euler sum usually agrees with the $\mathfrak{T}$ definition of series whenever both are defined.

The Gamma Function The gamma function, denoted by the Greek capital letter Γ, is a generalization to the discrete factorial function. It is defined for all complex numbers except the negative integers and zero. It satisfies:

$$\Gamma(n) = \int_0^{\infty} t^{n-1} e^{-t} \, dt = \frac{1}{n} \prod_{k=1}^{\infty} \frac{\left(1 + \frac{1}{k}\right)^n}{1 + \frac{n}{k}} = (n - 1)!$$

Both the gamma function and its infinite product representation are due to Euler.

The Harmonic Numbers The harmonic numbers H_n are given by:

$$H_n = \sum_{k=0}^{n-1} \frac{1}{1 + k}$$

If we let $\psi(n)$ denotes the digamma function, where $\psi = \frac{d}{dx}\log\Gamma(x)$, then harmonic numbers can be extended to the complex plane using:

$$H_n = \psi(n+1) + \gamma$$

Here, $\gamma \approx 0.5772$ is Euler's constant. Asymptotically, we have:

$$H_n \sim \gamma + \log n,$$

with an error that goes to zero as $n \to \infty$.

The Mittag-Leffler Star The Mittag-Leffler star of a function $f(x)$ around the origin is the set of all points $z \in \mathbb{C}$ such that the line segment $[0, z]$ does not pass through a singularity point of $f(x)$. For example, the star of the function $\log(1+x)$ includes the entire complex plane $\mathbb{C}$ except the negative real line $(-\infty, -1]$.

The Nörlund Means The Nörlund means are a generalization to the Cesàro means. Here, suppose p_j is a sequence of positive terms that satisfy $\frac{p_n}{\sum_{k=0}^n p_k} \to 0$. Then, the Nörlund mean of a sequence $(s_0, s_1, \ldots)$ is given by:

$$\lim_{n\to\infty} \frac{p_n s_0 + p_{n-1} s_1 + \cdots + p_0 s_n}{\sum_{k=0}^n p_k}$$

The limit of an infinite sequence $(s_0, s_1, \ldots)$ is defined by its Nörlund mean. All Nörlund means are regular, linear, and stable.

The Sampling Theorem The Sampling Theorem is a fundamental result in information theory. It states that band-limited functions with bandwidth B can be perfectly reconstructed from their discrete samples if the sampling rate is larger than $2B$. The requirement that the sampling rate exceeds twice the bandwidth is called the Nyquist criteria.

The $\mathfrak{T}$ Definition of Series The $\mathfrak{T}$ definition is a generalization to the classical definition of series. Given a series $\sum_{k=0}^\infty g(k)$, let:

$$h(z) = \sum_{k=0}^\infty g(k)\, z^k \tag{1}$$

In other words, $h(z)$ is the function whose Taylor series expansion around the origin is given by Eq. 1. If $h(z)$ is analytic in the domain $z \in [0, 1]$, then the $\mathfrak{T}$ value of the series $\sum_{k=0}^\infty a_k$ is defined by $h(1)$. For example, the infinite sum $1 - 2 + 3 - 4 + \ldots$ is assigned the value $\frac{1}{4}$ by $\mathfrak{T}$ because it arises out of the Taylor series expansion of $f(x) = (1+x)^{-2}$ at $x = 1$. The definition $\mathfrak{T}$ is regular, linear, and stable.

The $\mathfrak{T}$ Sequence Limit The $\mathfrak{T}$ sequence limit of an infinite sequence $S = (s_0, s_1, s_2, \ldots)$ is defined by the $\mathfrak{T}$-value of the infinite sum: $s_0 + \sum_{k=0}^\infty \Delta s_k$.

The χ Summability Method The summability method χ, which we introduced in this monograph, assigns to a series $\sum_{k=0}^{\infty} \alpha_k$ the value:

$$\sum_{k=0}^{\infty} \alpha_k \triangleq \lim_{n \to \infty} \sum_{k=0}^{n} \chi_n(k)\, \alpha_k, \qquad \text{where } \chi_n(k) = \prod_{j=1}^{k} \left(1 - \frac{j-1}{n}\right)$$

Triangular Numbers Triangular numbers is the sequence of integers $1, 3, 6, 10, \ldots$ whose kth element is given by $k(k+1)/2$. They are generated by the simple finite sum $\sum_{k=0}^{n-1}(k+1)$.

Index

χ summability method, 16, 17, 47, 50, 75, 98, 103, 120, 124, 128, 133, 136, 161
$\mathfrak{T}$ sequence limit, 70, 160. *see also* generalized definition $\mathfrak{T}$

Abel summability method, 12, 71, 75, 155
Academia Algebrae, 28
alternating power sum, 97
alternating sum, 94, 97, 98, 100
analytic continuation, 1, 49, 72, 99, 144
Aryabhata, 14
Aryabhatiya, 14
asymptotic analysis, 15, 42, 47, 58, 93, 98, 106, 125, 142
asymptotic differentiation order, 47, 97, 105, 116, 121, 124, 125, 127, 129, 156

backward difference operator, 134
bandlimited function, 137
bandwidth, 137–139
Barnes G-function, 14. *see also* superfactorial
Berndt-Schoenfeld periodic analogs, 15, 93
Bernoulli numbers, 4, 24, 60, 73, 96, 155, 157
Bernoulli-Faulhaber formula, 4, 8, 27, 49, 97, 116, 125
beta function, 39
Bohr-Mollerup theorem, 4, 36, 159
Boole summation formula, 15, 93, 95

Cauchy numbers, 147
Cauchy product, 68
Cauchy repeated integration formula, 58

ceiling function, 116
Cesàre means, 11, 16, 66, 71, 103, 159
Cesàro summability method. *see* Cesàro means
chain rule, 56
composite finite product, 12, 55
composite finite sum, 12, 55, 100, 129, 155
convergence acceleration, 98, 106, 124
cook-book recipe, 28. *see also* power sum

DFT. *see* discrete Fourier transform
digamma function, 13, 37, 146, 160. *see also* harmonic numbers
discrete Fourier transform, 106
discrete function, 5, 13
discrete harmonic function, 116
divergent series, 10, 16, 67, 93. *see also* summability theory
double gamma function, 14. *see also* superfactorial

Euler constant, 6, 30, 35, 37, 124, 146, 160
Euler polynomials, 95
Euler summation, 74, 99, 100, 159
Euler's generalized constants, 30
Euler-Macluarin summation formula, 15, 45, 47, 50, 93, 117, 124, 141

factorial function, 1–4, 6, 13, 16, 28, 35–37, 39, 42, 59, 89, 119, 124, 159. *see also* gamma function
falling factorial, 141
finite differences, 15, 133, 134, 143

© Springer International Publishing AG 2018
I. M. Alabdulmohsin, *Summability Calculus*,
https://doi.org/10.1007/978-3-319-74648-7

Printed in the United States
By Bookmasters